SpringerBriefs in Educational Communications and Technology

Published in collaboration with the AECT (Association for Educational Communications and Technology), Springer Briefs in Educational Communications and Technology focuses on topics of keen current interest in the broad area of educational information science and technology. Each Brief is intended to provide an introduction to a focused area of educational information science and technology, giving an overview of theories, issues, core concepts and/or key literature in a particular field. A brief could also provide:

- A timely report of state-of-the art analytical techniques and instruments in the field of educational information science and technology,
- A presentation of core concepts,
- An overview of a testing and evaluation method,
- A snapshot of a hot or emerging topic or policy change,
- An in-depth case study,
- A literature review,
- A report/review study of a survey, or
- An elaborated conceptual framework on model pertinent to educational information science and technology. The intended audience for Educational Communications and Technology is researchers, graduate students and professional practitioners working in the general area of educational information science and technology; this includes but is not limited to academics in colleges of education and information studies, educational researchers, instructional designers, media specialists, teachers, technology coordinators and integrators, and training professionals.

More information about this series at http://www.springer.com/series/11821

Aklilu Tilahun Tadesse • Pål Ingebright Davidsen
Erling Moxnes

Adapting Interactive Learning Environments to Student Competences

The Case for Complex Dynamic Systems

Springer

ASSOCIATION FOR EDUCATIONAL COMMUNICATIONS & TECHNOLOGY

Aklilu Tilahun Tadesse
Department of Geography
System Dynamics Group
University of Bergen
Bergen, Norway

Pål Ingebright Davidsen
Department of Geography
System Dynamics Group
University of Bergen
Bergen, Norway

Erling Moxnes
Department of Geography
System Dynamics Group
University of Bergen
Bergen, Norway

ISSN 2196-498X ISSN 2196-4998 (electronic)
SpringerBriefs in Educational Communications and Technology
ISBN 978-3-030-88288-4 ISBN 978-3-030-88289-1 (eBook)
https://doi.org/10.1007/978-3-030-88289-1

This Springer imprint is published by the registered company Springer Nature Switzerland AG
The registered company address is: Gewerbestrasse 11, 6330 Cham, Switzerland

Preface

This monograph focuses on the design of a personalized and adaptive online interactive learning environment (OILE) to enhance students' learning in and about complex dynamic systems (CDS). The study is motivated by research showing that most people, even experts, have difficulties comprehending CDS and communicating their understanding about such systems. The difficulties are due to challenges originating from (1) the structural complexity of CDS; (2) the skills required to produce and to infer dynamic behavior from the underlying systems structure; and (3) the effectiveness of methods, techniques, and tools that are available to us in our analysis of such systems.

While numerous studies have revealed the challenges people face when dealing with CDS, there are significant gaps in our understanding of how to improve cognitive and communicative capabilities in and about CDS and also of how to measure improvements. In this monograph, we provide some answers as to how we may best improve our cognitive capabilities to meet these challenges by way of effective instructional methods, techniques, and tools and their implementation in the form of an OILE.

The OILE developed for this purpose, builds on a five-step holistic instructional design framework; identification of instructional design models, identification of authentic learning material, identification of instructional methods, identification of instructional techniques, and design of the interface and implementation of the tool. In this OILE development, six well-documented instructional design models were considered; a four-component instructional design, first principles of instruction, constructivists learning environment, task-centered instruction, cognitive apprenticeship, and elaboration theory. The resulting OILE has the following three characteristics:

1. It presents an authentic, complex dynamic problem that the learner should address in its entirety. It then proceeds to allow learners to progress through a sequence of gradually more complex learning tasks.
2. It allows the learner to interact with the OILE while solving the problem at hand. Upon completion of each learning task and based on their individual performance,

the OILE provides the learners with information intended to facilitate the learning process. The support fades away as learners gain expertise.
3. It tracks and collects information on students' progress and generates learning analytics that are being used to assess students' learning and to tailor the information feedback to the students.

Overall, in this monograph, we discuss exhaustively the challenges associated with learning in and about CDS; present the theoretical design framework used in the development of the OILE; provide evaluation reports, a survey study and two impact studies to assess the design and implementation of the OILE using System Dynamics master's program students as subjects; discuss important lessons learned from the application of the design framework as a foundation for the OILE development; and finally, suggest recommendations for future studies.

We hope readers would find this monograph well suited to serve as a reading material in educational technology classes that address the planning, design, development, implementation, and evaluation of instructional materials. Because the monograph presents the design of OILEs starting from early planning and design, through an initial to a focused impact study taking into account learner differences and emphasizing the provision of individualized instructional scaffolding. Furthermore, the monograph accounts the three classical findings of research on learning; first that prior performance tends to predict future performance, which implies the need for diversity in scaffolding and feedback for low- and high-performing students; secondly that timely and informative feedback tends to enhance performance; and thirdly that time on task tends to predict performance. This monograph is among few studies that address all three of those major research findings in education.

In this monograph, we tried to address a difficult dual challenge in a higher education program. On the one hand, there is the inherent challenge of understanding complex and dynamic systems, both with educators and learners. On the other hand, there is the challenge of understanding learners with different backgrounds working in such a domain, that is, which challenges individual learners would face in different stages throughout their learning process. It is our belief that educators and educational technology students, who work in such a domain, can use the monograph as a reference to develop their own instructional materials.

Bergen, Norway

Aklilu Tilahun Tadesse
Pål Ingebright Davidsen
Erling Moxnes

Acknowledgments

This monograph is a result of a strong collaboration between a former PhD student and his two supervisors at the University of Bergen, System Dynamics group, Norway, that continued to this day. We would like to take this opportunity to acknowledge this strong collaboration. At the same time, we also would like to thank the System Dynamics master's program students, who participated in this research. Finally, we would like to extend our greatest appreciation to Prof. J. Michael Spector and his colleagues, who encouraged us to submit our work to the AECT books and briefs series.

Bergen, Norway

Aklilu Tilahun Tadesse
Pål Ingebright Davidsen
Erling Moxnes

Contents

Abbreviations

4C/ID	Four component instructional design
CDS	Complex dynamic systems
CLE	Constructivist learning environment
DBR	Design-based research
EF	Elaborated feedback
KCR	Knowledge of the correct response
KP	Knowledge of performance
KR	Knowledge of result/response
MCQ	Multiple-choice questions
MOOC	Massive open online course
OECD	Organization for Economic Co-operation and Development
OEQ	Open-ended questions
OILE	Online interactive learning environment
SD	System dynamics
TCI	Task-centered instruction

Chapter 1
Introduction

> *A problem arises when a [person] has a goal but does not know how... to [achieve it].*
> (Duncker, 1945, p. 1)

> *Problem-solving competence is an individual's capacity to engage in cognitive processing to understand and resolve problem situations where a method of solution is not immediately obvious. It includes the willingness to engage with such situations in order to achieve one's potential as a constructive and reflective citizen.*
> (OECD, 2013, p. 122).

> *Governments are increasingly confronted by uncertain and complex challenges whose scale and nature call for new approaches to problem solving. Some governments have started to use systems approaches in policy making and service delivery to tackle complex or "wicked" problems in areas ranging from education to ageing, healthcare and mobility. Systems approaches refer to a set of processes, methods and practices that aim to effect systems change.*
> (OECD, 2017, p.12)

> For students to succeed in work and life in the 21st century, one of the critical skills they should acquire is the ability to "*analyze how parts of a whole interact with each other to produce overall outcomes in complex systems*". The 21st century learning environments should "*enable students to learn in relevant, real-world 21st century contexts (e.g., through project-based or other applied work)*".
> (Battelle for Kids, 2019, pp. 4–8, in Partnership for Twenty-First Century Learning)

The quotes above highlight the general theme of this monograph, which is enhancing students' problem-solving competencies in a complex domain by designing learning environments that support and facilitate their learning. This chapter of the monograph first presents the main problem the research focuses on. It then provides the overarching purpose, research question and design of the research. The last section of the chapter defines the key terms used in the monograph.

A. T. Tadesse et al., *Adapting Interactive Learning Environments to Student Competences*, SpringerBriefs in Educational Communications and Technology,
https://doi.org/10.1007/978-3-030-88289-1_1

1.1 Need to Develop Complex Problem-Solving Skills

The past decade has seen a significant emphasis on the need for the development of twenty-first century skills, particularly on the importance of developing problem-solving skills in the complex domain (Griffin et al., 2012; Voogt et al., 2018). Problem-solving is not a new concept of the twenty-first century. As Karl Popper (1999) argues 'all life is problem-solving'. However, the ever-increasing complexity of the problems we face in both the public and private sector and their ever-changing nature signify the importance of increasing problem-solving skills.

Problems such as climate change, natural resource management issues, migration, famine, unemployment, healthcare issues etc. create significant challenges for both private and public organizations and threaten our survival (Sterman, 1994; Davidsen, 1996; Jonassen, 1997; Moxnes, 1998; Barlas, 2007; Griffin et al., 2012; OECD, 2017). These problems are often dynamic (i.e. develop over time) and they commonly originate from the internal structure of the systems with which the problems addressed are associated (Diehl & Sterman, 1995; Davidsen, 1996). The structure of a system is made up of the cause and effect relationships that exist between the attributes (variables) that define a system. And the complexity of a system is defined by the diversity of that system's structure.

A large body of studies shows that most people, even experts, have difficulties comprehending complex, dynamic systems (CDS) and communicating their understanding about such systems (Dörner, 1996; Moxnes, 1998, 2004; Cronin et al., 2009). These difficulties arise from limitations in three different types of capabilities: (1) The cognitive capability to comprehend structural complexity. (2) The skills required to infer the dynamic behavior of a system from its underlying structures. (3) The effectiveness of methods, techniques, and tools that are available to us in our analysis of such systems (Sterman, 1989; Davidsen, 1996; Spector & Anderson, 2000; Jonassen, 2000; Sawicka & Rydzak, 2007; Kopainsky et al., 2015; Ifenthaler & Eseryel, 2013; van Merriënboer & Kirschner, 2017).

Numerous studies demonstrate the challenges people face when dealing with CDS. However, there are significant gaps in our understanding of how to support and improve cognitive and communicative capabilities in and about CDS and also of how to measure the improvements or lack of improvements. John Sterman (2000), in his text book 'Business Dynamics', considers these challenges as among the 'major challenges in the further development of the field of System Dynamics' and calls for researchers to investigate the "type of experiences and education [that] might mitigate them and develop our systems thinking capabilities" (p. 896).

Learning theories suggest that to understand and communicate our understanding about CDS, we must develop adequate mental models about such systems (Seel, 2003; Kopainsky et al., 2015). One way for developing such mental models is using and/or building simulation models (Alessi, 2000; Sterman, 2000; Tennyson & Breuer, 2002; Seel, 2003). However, studies show that unless simulation models are accompanied with 'instructional overlays' such as interfaces that provide guidance, feedback, and tools to support learning, they would be insufficient to help learners build adequate and correct mental models (Spector & Davidsen, 1997; Alessi, 2000; Kopainsky et al., 2015).

Various efforts were made to design learning environments that support and facilitate students' learning while they are using and/or building simulation models (see for example Milrad et al., 2003; de Jong & van Joolingen, 2007; Pavlov et al., 2015). However, findings about the effectiveness of the learning environments in supporting and facilitating the development of adequate and correct mental models are mixed and inconsistent (Sawicka & Rydzak, 2007; Kopainsky et al., 2015).

In this research, we aimed at developing a design framework for personalized and adaptive online interactive learning environments (OILE) to enhance students' learning in and about complex dynamics systems. The design framework enabled the creation of an OILE with the following three features: (a) The OILE presents an authentic, complex dynamic problem that the learner should address in its entirety. It then proceeds to allow learners to progress through a sequence of gradually more complex learning tasks. (b) It allows for the learner to interact with the OILE while solving the problem at hand. Upon the completion of each learning task and based on their individual performance, the OILE provides the learners with information intended to facilitate the learning process. The support fades away as learners gain expertise. (c) The OILE tracks and collects information on students' progress. The information is used to tailor the feedback to the students while they are working on the OILE, and also to generate learning analytics that are being used to assess students' learning.

1.2 Purpose, Research Questions, and Design

The overarching purpose of this research is enhancing students' learning in and about complex, dynamic systems using personalized and adaptive online interactive learning environments. The main research question investigated in this monograph is:

> *How may we design personalized and adaptive OILEs that effectively enhance students learning in and about CDS?*

In order to investigate this research question further, it has been divided into sub-questions and examined under three different studies. Table 1.1 gives an overview of the three studies.

The overarching research design of the monograph is design-based research (DBR). Using DBR, real-world educational problems are analyzed, interventions are conceptualized, and then implemented iteratively in naturalistic settings. The objective is to produce new theories, artifacts, and practices that account for and potentially impact learning and teaching (Barab & Squire, 2004; Huang et al., 2019). Although DBR is difficult to conduct compared to other types of research, Herrington et al. (2007) encourage doctoral students particularly those who do educational technology research to utilize it as their main research design.

Table 1.1 Overview of the research

Research Purpose	To enhance students' learning in and about complex, dynamic systems using personalized and adaptive online interactive learning environments		
Main research question	How may we design personalized and adaptive OILEs that effectively enhance students learning in and about CDS?		
Studies	**Study I**	**Study II**	**Study III**
Research question	*How may one design online interactive learning environments to support individual students learning in and about complex dynamic systems?*	*Does using the Mr. Wang online interactive learning environment affect the development of students' complex dynamic problem-solving skills?*	*How does scaffolding feedback, which is integrated to the Mr. Wang OILE, affect the performance of students?*
Design	Design-based research Literature review Focus group discussion Survey	Design-based research Mixed methods research—Focus group discussion Single subject experiment Quasi-experiment	Design-based research Mixed methods research—Focus group discussion Single subject experiment Quasi-experiment
Sample	Peer-reviewed articles Previous literature Three cohorts of System dynamics master program students at the University of Bergen	Three cohorts of System dynamics master program students at the University of Bergen	Three cohorts of System dynamics master program students at the University of Bergen
Data	Inclusion/exclusion criteria Questionnaires	Assignments Demographic data Process log Posttest	Assignments Demographic data Process log Transfer skill exercises
Data analysis framework	Thematic analysis Coding/categorization Wilcoxon signed-Ranks test	Coding/categorization Paired samples t-tests Independent samples t-tests Effect size	Coding/categorization Paired samples t-tests Independent samples t-tests Person product-moment correlation coefficient Effect size

The primary advantage of DBR, according to Reeves (2006), is that it requires the research to pass through at least four steps. The four steps, which have been incorporated in the different parts of this research, are: (1) Analysis of practical teaching-learning problem through collaboration between researchers and practitioners (Studies I and II). (2) Creation of prototype solutions based on existing design principles (Studies I, II, and III). (3) Testing and refinement of solutions in iterative cycles (Studies I, II, and III). (4) Reflection and production of design principles (Studies I, II, and III).

DBR naturally leads to the use of mixed methods research combining qualitative and quantitative methods (Anderson & Shattuck, 2012; McKenney & Reeves, 2013; Creswell & Creswell, 2018). Table 1.1 summarizes the different research methods applied in the three studies; literature review, focus group discussion, survey, single subject, and quasi-experiments.

1.3 Definition of Key Terms

This section presents some of the key terms used in this monograph, which we adopted from the literature.

Learning is a "stable and persistent changes in what a person knows, believes, and/or can do" (Spector, 2017, p. 1421). In the context of this monograph, the broad definition of learning given by Spector includes the positive change in the five major categories of learning identified by Gagné (1985): verbal information, intellectual skills, cognitive strategies, motor skills, and attitudes. Hence, when we refer to what a person knows, believes and/or can do, we are referring to Gagné's five major categories of learning.

Instruction "is that which is designed and/or intended to support, enhance, or improve learning" (Spector, 2018, p. 35). It is a 'deliberate' and 'goal-directed' activity (Merrill, 2013).

Instructional design is a systematic "planning, selection, sequencing, and development of activities and resources to support targeted learning outcomes" (Spector, 2015, p. 221).

Personalized learning is "the dynamic configuration of learning activities, assignments, and resources to fit individual needs and expectations, based on an automated analysis of student profiles, past performance, current learning needs and difficulties, and what has worked for similar students with similar learning needs and difficulties" (Spector, 2015, 223).

Interaction is "the give and take between one or more learners and an instructional system or environment that may include human tutors and teachers as well as technology facilitated components" (Spector, 2015, p. 222).

Learning environment is a "specific arrangement or setting for teaching and learning [to occur]" (Seel et al., 2017, p. 4).

Adaptive learning environment is a learning environment that "aims at supporting learners in acquiring knowledge and skills in a particular learning domain. The goal is to enhance the individual learning process with respect to speed, accuracy, quality and quantity of learning" (Weber, 2012, p. A113).

In the monograph, the learning environment is designed targeting individual students need and is planned to occur in a digital (online) platform through a continuous interaction between individual learners and the learning environment. Thus, the name personalized and adaptive online interactive learning environment (OILE) has been used in this monograph to refer such a teaching-learning setting.

References

Alessi, S. (2000). Designing educational support in system-dynamics-based interactive learning environments. *Simulation & Gaming, 31*(2), 178–196.

Anderson, T., & Shattuck, J. (2012). Design-based research: A decade of progress in education research? *Educational Researcher, 41*(1), 16–25.

Barab, S. A., & Squire, K. (2004). Design-based research: Putting a stake in the ground. *The Journal of the Learning Sciences, 13*(1), 1–14.

Barlas, Y. (2007). System dynamics: Systemic feedback modeling for policy analysis. *System, 1*(59), 1–68.

Battelle for Kids. (2019). Framework for 21st century learning definitions. In *Partnership for 21st century learning*. Retrieved April 20, 2020, from http://static.battelleforkids.org/documents/p21/P21_Framework_DefinitionsBFK.pdf

Creswell, J. W., & Creswell, J. D. (2018). *Research design; Qualitative, quantitative, and mixed methods approaches* (5th ed.). Sage.

Cronin, M. A., Gonzalez, C., & Sterman, J. D. (2009). Why don't well-educated adults understand accumulation? A challenge to researchers, educators, and citizens. *Organizational Behavior and Human Decision Processes, 108*(1), 116–130.

Davidsen, P. I. (1996). Educational features of the system dynamics approach to modelling and simulation. *Journal of Structural Learning, 12*(4), 269–290.

de Jong, T., & van Joolingen, W. R. (2007). Model-facilitated learning. In J. M. Spector, M. D. Merrill, J. J. van Merriënboer, & M. P. Driscoll (Eds.), *Handbook of research on educational communications and technology* (pp. 457–468). Lawrence Erlbaum Associates: Taylor & Francis Group.

Diehl, E., & Sterman, J. D. (1995). Effects of feedback complexity on dynamic decision making. *Organizational Behavior and Human Decision Processes, 62*(2), 198–215.

Dörner, D. (1996). *The logic of failure: Recognizing and avoiding error in complex situations*, [translated by Rita and Kimber, R.]. Metropolitan Books.

Duncker, K. (1945). On problem-solving (L. S. Lees, Trans.). *Psychological Monographs, 58*(5), i–113.

Gagné, R. M. (1985). *The conditions of learning and theory of instruction* (4th ed.). Holt, Rinehart and Winston.

Griffin, P., McGaw, B., & Care, E. (Eds.). (2012). *Assessment and teaching of 21st century skills*. Springer.

Herrington, J., McKenney, S., Reeves, T. C., & Oliver, R. (2007). Design-based research and doctoral students: Guidelines for preparing a dissertation proposal. In *proceedings of world conference on educational multimedia, hypermedia and telecommunications* (pp. 4089–4097). AACE.

Huang, R., Spector, J. M., & Yang, J. (2019). Design-based research. In *educational technology. Lecture notes in educational technology*. Springer.

Ifenthaler, D., & Eseryel, D. (2013). Facilitating complex learning by mobile augmented reality learning environments. In R. Huang, S. Kinshuk, & J.M. (Eds.), *Reshaping learning: Frontiers of learning technology in a global context* (pp. 415–438). Springer.

Jonassen, D. H. (1997). Instructional design models for well-structured and ill-structured problem-solving learning outcomes. *Educational Technology Research and Development, 45*(1), 65–94.

Jonassen, D. H. (2000). Toward a design theory of problem solving. *Educational Technology Research and Development, 48*(4), 63–85.

Kopainsky, B., Alessi, S. M., Pedercini, M., & Davidsen, P. I. (2015). Effect of prior exploration as an instructional strategy for system dynamics. *Simulation & Gaming, 46*(3–4), 293–321.

McKenney, S., & Reeves, T. C. (2013). Systematic review of design-based research progress: Is a little knowledge a dangerous thing? *Educational Researcher, 42*(2), 97–100.

Merrill, M. D. (2013). *First principles of instruction: Identifying and designing effective, efficient and engaging instruction*. Pfeiffer.

Milrad, M., Spector, M., & Davidsen, P. (2003). Model facilitated learning. In S. Naidu (Ed.), *Learning and teaching with technology: Principles and practices* (pp. 11–24). Kogan Page.
Moxnes, E. (1998). Not only the tragedy of the commons, misperceptions of bioeconomics. *Management Science, 44*(9), 234–1248.
Moxnes, E. (2004). Misperceptions of basic dynamics, the case of renewable resource management. *System Dynamics Review, 20*(2), 139–162.
OECD. (2013). *PISA 2012 Assessment and analytical framework: Mathematics, reading, science, problem solving and financial literacy*. Retrieved May 28, 2016, from http://www.oecd.org/pisa/pisaproducts/PISA%202012%20framework%20e-book_final.pdf
OECD. (2017). *Systems approaches to public sector challenges: Working with change*. OECD Publishing.
Pavlov, O. V., Saeed, K., & Robinson, L. W. (2015). Improving instructional simulation with structural debriefing. *Simulation & Gaming, 46*(3–4), 383–403.
Popper, K. R. (1999). *All life is problem solving* (translated by Patrick Camiller). Routledge, Taylor & Francis.
Reeves, T. C. (2006). Design research from a technology perspective. In J. J. H. van den Akker, K. Gravemeijer, S. McKenney, & N. Nieveen (Eds.), *Educational design research* (pp. 52–66). Routledge.
Sawicka, A., & Rydzak, F. (2007). Incorporating delays in the decision-making interface: An experimental study. In *Paper presented at the 25th international conference of the system dynamics society*. MA.
Seel, N. M. (2003). Model-centered learning and instruction. *Technology, Instruction, Cognition and Learning, 1*(1), 59–85.
Seel, N. M., Lehmann, T., Blumschein, P., & Podolskiy, O. A. (2017). *Instructional design for learning: Theoretical foundations*. Springer.
Spector, J. M. (2015). *Foundations of educational technology: Integrative approaches and interdisciplinary perspectives* (2nd ed.). Routledge.
Spector, M. (2017). Reflections on educational technology research and development. *Educational Technology Research and Development, 64*, 1415–1423.
Spector, J. M. (2018). Smart learning environments: Potential and pitfalls. In K. Persichitte, A. Suparman, & M. Spector (Eds.), *Educational technology to improve quality and access on a global scale* (pp. 33–42). Springer.
Spector, J. M., & Anderson, T. M. (2000). *Integrated and holistic perspectives on learning, instruction and technology*. Kluwer Academic Publishers.
Spector, J. M., & Davidsen, P. I. (1997). Creating engaging courseware using system dynamics. *Computers in Human Behavior, 13*, 127–155.
Sterman, J. D. (1989). Misperceptions of feedback in dynamic decision making. *Organizational Behavior and Human Decision Processes, 43*(3), 301–335.
Sterman, J. D. (1994). Learning in and about complex systems. *System Dynamics Review, 10*(2–3), 291–330.
Sterman, J. D. (2000). *Business dynamics: Systems thinking and modeling for a complex world*. Irwin McGraw-Hill.
Tennyson, R. D., & Breuer, K. (2002). Improving problem solving and creativity through use of complex-dynamic simulations. *Computers in Human Behavior, 18*(6), 650–668.
van Merriënboer, J. J., & Kirschner, P. A. (2017). *Ten steps to complex learning: A systematic approach to four-component instructional design*. Routledge.
Voogt, J., Knezek, G., Christensen, R., & Lai, K. (Eds.). (2018). *Second handbook of information technology in primary and secondary education*. Springer International Handbooks of Education.
Weber, G. (2012). Adaptive learning systems. In N. M. Seel (Ed.), *Encyclopedia of the sciences of learning* (pp. A113–A115). Springer.

Chapter 2
Challenges with Supporting Learning in and about Complex Dynamic Systems

We experience a multifaceted challenge when we try to understand complex, dynamic systems and communicate our understanding about such systems. Figure 2.1 summarizes the different layers of difficulties we experience while dealing with CDS. At the core of the challenges lies the subject matter, the structural complexity of dynamic systems (red colored circles). Then comes us—the people who try to understand such system and communicate our understanding to others (orange colored circles). At the top of the layers are the theories, methods, techniques, and tools and their level of effectiveness in supporting and measuring our understanding of the core subject matter. This section gives an overview of the challenges in each layer, how the different layers synergize to make our problems more complex, and efforts made to address the challenges.

2.1 Structural Complexity of CDS and Challenges with Understanding Them

The problems we face often have a dynamic nature and commonly originate from the internal structure of the system that generates the problem (Sterman, 1994; Diehl & Sterman, 1995; Davidsen, 1996). We are cognitively challenged when investigating how dynamics (change over time) develops based on the underlying structure (Moxnes & Saysel, 2009).

The structure of a dynamic system is often characterized by a set of accumulation processes, all of which are interrelated by way of causal, non-linear feedback. Such a structure gives rise to the generation of rich patterns of behavior. The difficulty arises in understanding and conveying our understanding of how this structural complexity generates the associated behavior and, most importantly, how that behavior feeds back to the structure and shifts endogenously the relative significance (dominance) of the structural components over time. That is, certain

A. T. Tadesse et al., *Adapting Interactive Learning Environments to Student Competences*, SpringerBriefs in Educational Communications and Technology,
https://doi.org/10.1007/978-3-030-88289-1_2

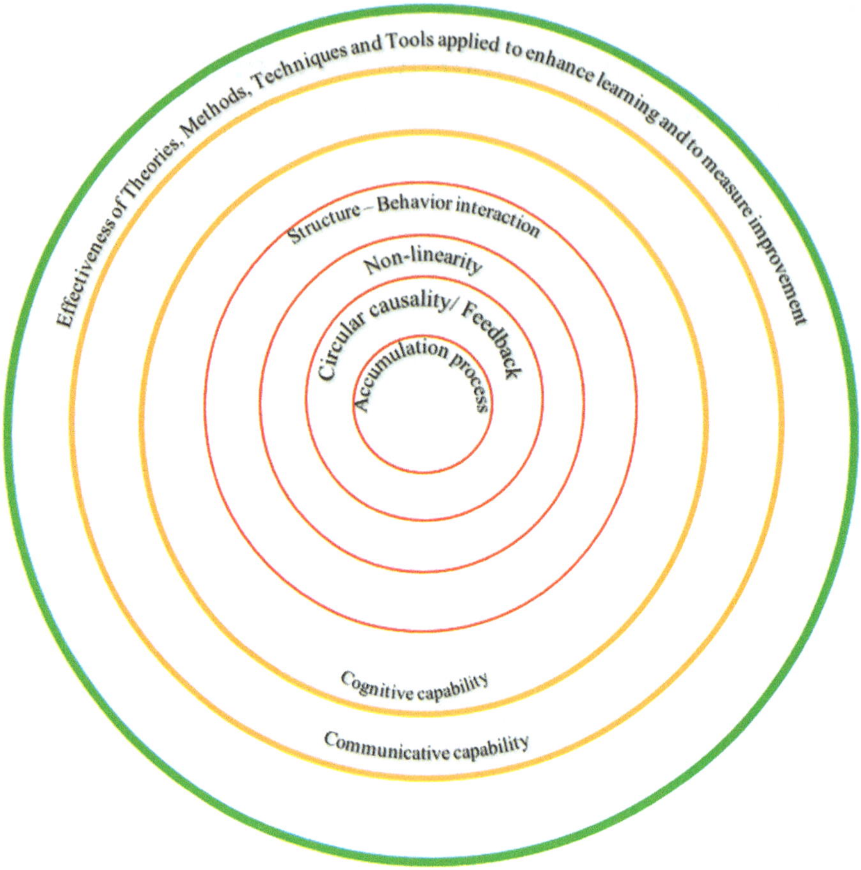

Fig. 2.1 Layers of challenges in and about complex dynamic systems

behavioral responses of the system may activate dormant feedback loops and make them dominant in the system. Unless we do proper analysis, we often fail to recognize these dormant feedback loops, which may dominate the system later (Davidsen, 1996).

The first difficulty arises from understanding what kind of behavior the accumulation process generates. The accumulation process comprises at least three elements—a stock (an accumulator), a flow (one that increases or drains the stock), and a delay (time lag). During the accumulation process, the behavior of the flow is transformed into the behavior of a stock, which does not have the same shape as the flow behavior, and it does so only as time progresses, that is, with a time lag/delay (Diehl & Sterman, 1995; Davidsen, 1996; Moxnes & Jensen, 2009; Cronin et al., 2009). Thus, if we cannot observe the accumulation process (i.e. the systems structure) that transforms one behavior (that of the flow) to another behavior (that of the stock), we face a considerable challenge when trying to understand the dynamics of the system.

The second difficulty arises from understanding and conveying our understanding of the behavior generated by circular causality (that is, a feedback loop). That requires circular reasoning that, in static systems, implies simultaneity that, in dynamic systems, can only be resolved with the progression of time (by intervening accumulation processes). In other words, the circular reasoning is elevated by the fact that we need not only to go around the circle, but also to progress in time. Consider a system with two elements, say A and B. Assume that a feedback loop in this system relates A to B then feeds back to A. In a static system, a change in A instantaneously affects B. If B instantaneously influences A, there is simultaneity. In a dynamic system, a change in A may cause an instantaneous effect in B. However, for a change in B then to affect A, a certain amount of time needs to pass, i.e. an accumulation process must take place.

The third difficulty arises when we try to understand behavior resulting from non-linearity. In non-linear systems, feedback loops synergize to impact the systems (Davidsen, 1996). Hence, the main challenge is that the effect of one cause is conditioned by the size of another cause (Sterman, 1994; Davidsen, 1996; Barlas, 2007). Thus, to understand the effect, one must keep two causes in mind at the same time. However, our cognitive capacity limits us from doing so. Cognitive psychology have found that people struggle and use biased heuristics such as anchoring and adjustment (Tversky & Kahneman, 1974).

Hence, as we go from accumulation via feedback to non-linearity, the individual complexities do not merely add up but synergize and make it very challenging to understand and convey explanations of the behavior of CDS.

Our challenge in understanding and reasoning about CDS is further amplified because the behavior generated feeds back to the structure and shifts endogenously the relative significance of the structural components over time; that is, it may activate dormant feedback loops (Davidsen, 1996; Sterman, 2002). Initially, these dormant loops are not likely to be considered by most people including many analysts. However, proper analysis requires that dormant feedback loops are considered before they start to dominate behavior. Sterman (2002) exemplifies this fourth cognitive difficulty with some real-world issues:

> "All too often, well-intentioned efforts to solve pressing problems create unanticipated side effects. Our decisions provoke reactions we did not foresee ... and in a number of occasions, the responses of the systems to the interventions defeat the intervention themselves. From California's failed electricity reforms in USA, to road building programs that create suburban sprawl and actually increase traffic congestion, to pathogens that evolve resistance to antibiotics, our best efforts to solve problems often make them worse" (p. 2).

All these difficulties pose severe learning challenges when facing CDS. In order to facilitate learning about CDS, we need to equip ourselves with proper teaching strategies supported by educational theories, methods, techniques, and tools that meet the structural complexity underlying such cognitive challenges (Sterman, 1994; Davidsen, 1996). There are numerous studies that demonstrate the challenges people face when dealing with CDS. However, there are gaps in our understanding of how to improve as well as on how to measure the improvement or lack of improvement in students' capability of understanding and communicating their understanding about CDS.

2.2 Theories, Methods, Techniques, and Tools for Supporting Learning in and about CDS

The challenge with understanding and communicating our understanding about CDS is pervasive. However, there are new technologies that provide opportunities and affordance to support learning in and about CDS. Consequently, instructional designers and learning scientists are looking for best ways to support students' learning (Spector & Anderson, 2000). This section provides a brief review of the existing instructional design theories and/or models, methods, techniques, and tools used to support learning in and about CDS and highlights the gaps that still need further research.

2.2.1 Instructional Design Theories for Supporting Learning in and about CDS

It is quite common to see the phrase 'instructional design theories' and 'instructional design models' in instructional design literature. For example, Reigeluth (1999a) uses the phrase instructional design theory, whereas Seel et al. (2017) prefer 'instructional design models'. In this monograph, the two phrases have been used alternatively to refer the same concept, 'that which offers explicit guidance on how to better help people learn and develop'.

There are numerous instructional design theories and/or models in the literature that are proposed to foster the systems thinking and the holistic perspective, a perspective that claims *the whole is always more than the sum of its parts*—indicating that the structure of parts synergizes to produce the resulting dynamics of a system. During the selection or design of instructional design theories and models, according to Spector (2000), one should consider the following five basic principles:

- Learning Principle—learning is fundamentally about change.
- Experience Principle—experience is the starting point for understanding.
- Context Principle—context determines meaning.
- Integration Principle—relevant contexts are broad and multi-faceted.
- Uncertainty Principle—we know less than we are inclined to believe. (p. 524)

This research utilized these five basic principles and the holistic perspective as inclusion criteria for choosing instructional design theories and/or models to inform the design of a learning environment that enhance students' learning in and about CDS.

Six instructional design theories and models were considered in the research that fulfils the above criteria: The Four Component Instructional Design model (4C/ID, van Merriënboer & Kirschner, 2017), First Principles of Instruction (Merrill, 2002, 2013), Constructivist Learning Environments (CLE, Jonassen, 1999), Task Centered Instruction (TCI, Francom & Gardner, 2014; Francom, 2017), Cognitive Apprenticeship (Collins et al., 1989, 1991), and Elaboration Theory (Reigeluth, 1999b).

The primary emphasis regarding the design of learning environment varies across the six instructional design models. Nevertheless, the models have four key features that make them suitable for designing learning environments that foster understandings of CDS.

First, they offer a unifying perspective regarding learning tasks. These instructional design models argue that the learning tasks should;

- be at the center of the instructional design
- be based on authentic problems
- comprise the entire knowledge and skills that learners would be able to acquire when they complete the entire learning tasks
- be designed in a way that learners can address the authentic problem in its entirety, from "start to finish, rather than discrete pieces" of the problem
- be designed in a way that learners can progress from simple to complex steps in their analysis of the entire task

Second, the instructional design models underscore the importance of providing instructional support that gradually fades away over time as learners gain expertise.

Third, they all promote holistic instructional design (Spector & Anderson, 2000; van Merriënboer & Kirschner, 2017). They recognize the dynamic interdependency between the elements that constitute an instructional system of complex learning that makes the instructional system an irreducible whole.

Fourth, they emphasize the importance of transfer of knowledge and skills to everyday life. The tasks the learners undertake as part of the learning experience and the instruction they follow in the learning environment should help the learners transfer their knowledge and skills to related real world settings.

Despite there are numerous instructional design theories and/or models with detailed prescriptive instructions on 'how to better help people learn and develop', the actual practice of putting theory into practice is largely missing (Spector & Anderson, 2000). Seel et al. (2017) share Spector's view of the lack of putting theory into practice and expressed their hope how this situation might change:

> "Instructional design denominates an educational discipline that is concerned with the development of theories of effective teaching and learning as well as with their conversion into educational practice. [...However, in practice,] instructional design models mostly have been created on the round table of theorists, and in consequence, lack a systematic evaluation of their suitability for daily use. Hence, the quality of published research in the field of instructional design has been criticized in general as poor. [...An increase in the use

of] design-based research, [which] demonstrated considerable potential to advance design, research, and practice in the field of instructional design,...[however], may bridge the gap between theory and practice". (Seel et al., 2017, pp. 13–109)

This monograph provides evidence regarding how to put existing instructional design theories and/or models proposed to support learning in and about CDS into practice using design-based research as its overarching research design.

2.2.2 *Instructional Methods for Supporting Learning in and about CDS*

Reigeluth (1999a) argue that instruction design theories should "identify methods of instruction (ways to support and facilitate learning) and the situation in which those methods should and should not be used" (p. 6). He also argues that the identified "methods need to be broken into more detailed component methods to provide more guidance" (p. 7). In line with Reigeluth's argument, this research identified instructional methods from the existing literature that can facilitate the students' learning in and about CDS. The methods are further divided into 'detailed component methods', referred here as 'instructional techniques'. In this research, the techniques were manifested in the form of an educational tool, a learning environment that supports the students in their study of CDS. This subsection presents the instructional method identified from existing literature that can possibly support learning in and about CDS.

In the above subsection, we noted that the six instructional design models have four common feature that make them suitable for designing learning environments that support leaning in and about CDS. Of the four feature, one is the importance of providing instructional support that gradually fades away over time as learners gain expertise. Literature in the STEM (Science, Technology, Engineering, and Mathematics) fields show that instructional scaffolding method has been widely applied and is found very effective in offering instructional support that gradually fades away over time as learners gain expertise (Belland, 2017).

Wood et al. (1976) first introduced the instructional scaffolding method while they describe how a tutor provided support to children when the children were constructing pyramids with wooden blocks. Wood and his colleagues define scaffolding as a "process that enables a child or novice to solve a problem, carry out a task or achieve a goal which would be beyond his unassisted efforts" (Wood et al., 1976, p. 90). The support provided is "meant to extend students' current abilities" so that they can carry out the "bulk of the work required to solve the problem" (Belland, 2017, p. 17).

The instructional scaffolding method comprises three elements: dynamic assessment, provision of just the right amount of support, and intersubjectivity (Belland, 2017). The dynamic assessment determines whether the learners are constructing knowledge and skills from the learning tasks and whether they are on the right path to be able to perform the tasks independently. If the assessment indicates that the learners are having difficulties to make meaningful learning, the scaffold level increases to provide the right amount of support. If the learners are on the right path, the scaffold gradually fades away (Wood et al., 1976; Belland, 2017). Intersubjectivity refers to a shared understanding between the "scaffolder" (the teacher or the learning environment) and the "scaffoldee" (the learner) regarding a successful performance of a learning task (Belland, 2017). The research utilized the instructional scaffolding method to facilitate the students' learning and their development in their study in and about CDS.

2.2.3 Instructional Techniques for Supporting Learning in and about CDS

Reigeluth (1999a) emphasized that the instructional methods identified for supporting and facilitating learning need to be further broken into component methods, which has been referred to in this monograph as instructional techniques. As descried above, the instructional scaffolding method has three elements: dynamic assessment, provision of just the right amount of support, and intersubjectivity. In order to effectively support and facilitate the students learning, these three elements of the instructional scaffolding method need to be broken into instructional techniques.

Literature show that educational feedback is an instructional technique, where the three elements of the instructional scaffolding method can be manifested, and that has powerful influence on students learning, both positively and negatively (Sadler, 1989; Hattie & Timperley, 2007; Shute, 2008; Narciss, 2017). In an educational context, feedback is defined as an information communicated to a student about a gap in performance between actual and desired performance so as to alter the gap (Ramaprasad, 1983). Positive impacts were observed when students receive feedback either in synergy or separately on aspects of a task (an item) being undertaken and/or on how to do it more effectively. However, when students receive feedback only on aspects of personal attributes, for example, praise, rewards, and/or punishments, the observed impact is negative (Kluger & DeNisi, 1996; Hattie & Timperley, 2007; Shute, 2008).

Students can receive feedback from various sources of information such as from a teacher, a peer, a family member, a book or a programmed system such as online interactive learning environments. However, only few knowledge and skill sets can be acquired satisfactorily simply through being informed about them, most require practices in a supportive learning environment that involves an agent such as a teacher (Sadler, 1989). In such supportive learning environments, Sadler (1989) argues, the agent helps to (a) identify the skills that are to be learned, (b) recognize and describe a fine performance, (c) demonstrate a fine performance, and (d) indicate how a poor performance can be improved up until the stage, where the students in themselves can self-regulate and perform the items on their own.

Sadler's and other researchers view regarding the positive influence of educational feedback can be summarized with a causal loop diagram shown in Fig. 2.2, which is a System Dynamics representation for causal interaction between two or more variables. In the causal loop diagram, a "+" sign indicates an increase in the cause would result an increase in the effect and a "−" sign indicates an increase in the cause would result a decrease in the effect given everything else kept constant. The "R" indicates a reinforcing loop and the "B" indicates a counterbalancing loop. The double lines on the connecting arrow indicate the presence of a time lag (delay).

When an external agent notices a gap between the desired knowledge level a student should have and the student's knowledge level the agent perceived, she/he provides support to the student. The support would then increase the student's internal knowledge level about the material she/he is studying, possibly after some delay due to the time needed for assimilating the new knowledge. The increase in the student's internal knowledge level would then be reflected in her/his performance level on an assessment task, given the quality of the assessment instrument applied to measure the student's knowledge. This in turn provides a base for the external agent to either increase or reduce the support level for the student next time around.

In short, on the one hand, an increase in support level from the external agent increases the student's internal knowledge level, which increases the student's performance, which in turn increases the agent's perception of the student's knowledge level, which then force the support level to decrease (fade way) next time round. Thus, the larger circular loop in the diagram acts as a counteracting loop. On the

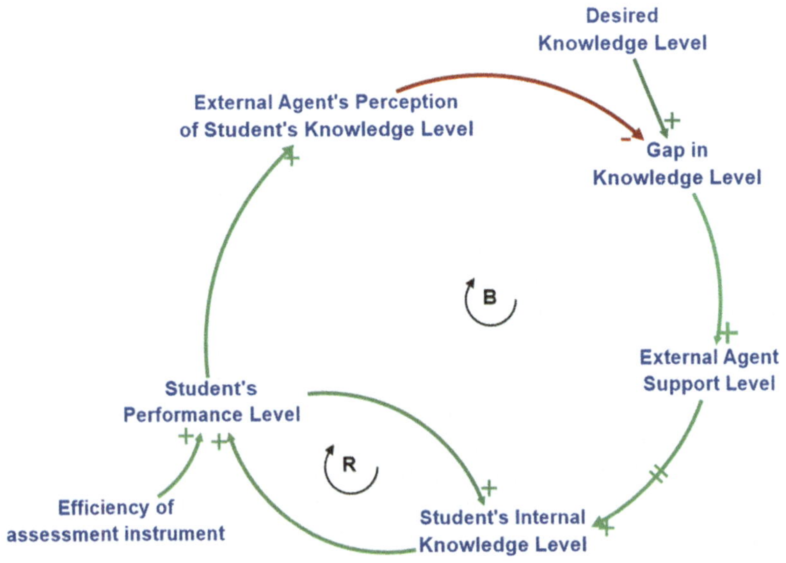

Fig. 2.2 Causal loop diagram for educational feedback

other hand, an increase in the student's internal knowledge level following the support from the external agent, increases the student's performance level—which in turn increases the student's internal knowledge level in a reinforcing manner. Hence, the smaller circular loop at the bottom of the diagram, after receiving the minimum support it needs to stand on its own, acts as a self-reinforcing loop allowing the student to build her/his own knowledge independent of the external agent.

Providing such a supportive learning environment with the human agent, particularly in higher education institutes, is often very difficult. This is due in part to the sheer increase in enrollment of students, difference in individual student's need, and shortage of the human capital that would be able to respond to the needs of each and every student (Bloom, 1984; Graesser et al., 2005; Carless et al., 2011; Narciss, 2008, 2013; Boud & Molloy, 2013). However, with the emergence of advanced computer based instructional technologies, designing OILE that effectively support students' learning is increasingly becoming possible albeit it is time consuming and is at its infancy (Graesser et al., 2005; Spector, 2009; Narciss, 2013; Van der Kleij et al., 2015; Kim & Ifenthaler, 2019). This research demonstrates how to effectively design and integrate an educational feedback called scaffolding feedback, which is a synergy of different feedback types that fade away as learners gain expertise, with the OILE to foster the students' learning in and about CDS. Also, the research shows how the three elements of the instructional scaffolding can be manifested in the form of scaffolding feedback in the OILE.

2.2.4 Instructional Tools for Supporting Learning in and about CDS

Designing interactive learning environments that effectively support learning in and about CDS is a challenging but fascinating task, which requires the synthesis of different instructional theories, methods, and techniques, which should be manifested in the form of an educational tool, the learning environment (Sterman, 1994; Davidsen, 2000; Eseryel et al., 2011). The task is difficult because these learning environments are required to influence the formation of mental models that govern learners' decision-making and action in CDS (Davidsen, 2000). Moreover, the learning environments are required to provide contexts to practice scientific methods in both virtual and real worlds, while facilitating the practice (Sterman, 1994). A review of the literature in the area of supporting learning in and about CDS show but a few interactive learning environments that have been designed to offer support for 'using' and/or 'building' computer models.

Students use computer simulations built by others either to conduct controlled experiments (trials and occasional success), as in the case of management flight simulators, or to gain training on specific tasks/procedures, for example, driving a car or flying an aircraft, without diving into the inner workings of the devices (Sterman, 1994; Alessi, 2000; Richardson, 2014; Pavlov et al., 2015). Learning environments that promote using simulations often have user-friendly interfaces that allow users to manipulate input variables and study changes in the output.

One of the drawbacks of such learning environments is that their simulators are 'black boxes' (Alessi, 2000). They provide little or no information about the underlying structure of the complex dynamic system (Alessi, 2000; de Jong & van Joolingen, 2007; Pavlov et al., 2015). The interfaces of such platforms often display surface relationship between input provided by the user and output provided by the simulation engine. Such platforms often provide open loop systems, where the users provide inputs hoping to change the state of the system under study in some desired way and the system reacts and provides output based on its invisible underlying structures. However, as Dörner (1996) signifies, understanding and solving complex and dynamic problems require the skill of providing causal and structural explanations for changes that happen in the system. It is unlikely that such a skill would be acquired only by manipulating input variables and viewing the changes in the systems behavior externally (Milrad et al., 2003). To acquire such a skill, perhaps the students need to be engaged in computer modeling activities (Sterman, 1994; Davidsen, 1996; Dörner, 1996; Alessi, 2000; Milrad et al., 2003).

The other drawback with learning environments that promote using simulations is that they do not require learners to pass through rigorous scientific methods such as problem identification, hypothesis formulation, carrying out analysis, and interpreting results. There are few other learning environments that have been designed to promote such learning, where learners are engaged in a scientific inquiry to uncover the underlying structure of CDS and solve their associated problems by building computer models (Alessi, 2000).

Students build/create computer simulations to develop deeper understanding about the underlying structures of CDS. Studies conducted in this domain, though they are scarce, show positive impact on the students' understanding of CDS (de Jong & van Joolingen, 2007). The remaining problem that needs additional research is how to design an effective learning environment that supports both the model using and model building features to improve the students' cognitive and communicative capabilities in and about CDS, while also accounting for the needs of individual students. This monograph presents an approach for doing so and reports results obtained from applying the approach.

References

Alessi, S. (2000). Designing educational support in system-dynamics-based interactive learning environments. *Simulation & Gaming, 31*(2), 178–196.

Barlas, Y. (2007). System dynamics: Systemic feedback modeling for policy analysis. *System, 1*(59), 1–68.

Belland, B. (2017). *Instructional scaffolding in STEM Education: Strategies and efficacy evidence*. Springer Open.

Bloom, B. S. (1984). The 2 sigma problem: The search for methods of group instruction as effective as one-to-one tutoring. *Educational Researcher, 13*(6), 4–16.

Boud, D., & Molloy, E. (2013). Rethinking models of feedback for learning: The challenge of design. *Assessment and Evaluation in Higher Education, 38*(6), 698–712.

Carless, D., Salter, D., Yang, M., & Lam, J. (2011). Developing sustainable feedback practices. *Studies in Higher Education, 36*(1), 395–407.

Collins, A. M., Brown, J. S., & Newman, S. E. (1989). Cognitive apprenticeship: Teaching the crafts of reading, writing, and mathematics. In L. B. Resnick (Ed.), *Knowing, learning, and instruction: Essays in honor of Robert Glaser* (pp. 453–494). Lawrence Erlbaum Associates.

Collins, A. M., Brown, J. S., & Holum, A. (1991). Cognitive apprenticeship: Making thinking visible. *American Educator, 15*(3), 6–11.

Cronin, M. A., Gonzalez, C., & Sterman, J. D. (2009). Why don't well-educated adults understand accumulation? A challenge to researchers, educators, and citizens. *Organizational Behavior and Human Decision Processes, 108*(1), 116–130.

Davidsen, P. I. (1996). Educational features of the system dynamics approach to modelling and simulation. *Journal of Structural Learning, 12*(4), 269–290.

Davidsen, P. I. (2000). Issues in the design and use of system-dynamics-based interactive learning environments. *Simulation & Gaming, 31*(2), 170–177.

de Jong, T., & van Joolingen, W. R. (2007). Model-facilitated learning. In J. M. Spector, M. D. Merrill, J. J. van Merriënboer, & M. P. Driscoll (Eds.), *Handbook of research on educational communications and technology* (pp. 457–468). Lawrence Erlbaum Associates: Taylor & Francis Group.

Diehl, E., & Sterman, J. D. (1995). Effects of feedback complexity on dynamic decision making. *Organizational Behavior and Human Decision Processes, 62*(2), 198–215.

Dörner, D. (1996). *The logic of failure: Recognizing and avoiding error in complex situations*, [translated by Rita and Kimber, R.]. Metropolitan Books.

Eseryel, D., Ge, X., Ifenthaler, D., & Law, V. (2011). Dynamic modeling as a cognitive regulation scaffold for developing complex problem-solving skills in an educational massively multiplayer online game environment. *Journal of Educational Computing Research, 45*(3), 265–286.

Francom, G. M. (2017). Principles for task-centered instruction. In C. M. Reigeluth, B. J. Beatty, & R. D. Myers (Eds.), *Instructional design theories and models: The learner-centered paradigm of education* (Vol. 4, pp. 65–91). Taylor & Francis.

Francom, G. M., & Gardner, J. (2014). What is task-centered learning? *TechTrends, 58*(5), 27–35.

Graesser, C., McNamara, D. S., & VanLehn, K. (2005). Scaffolding deep comprehension strategies through Point & Query, AutoTutor, and iSTART. *Educational Psychologist, 40*(4), 225–234.

Hattie, J., & Timperley, H. (2007). The power of feedback. *Review of Educational Research, 77*, 81–112.

Jonassen, D. H. (1999). Designing constructivist learning environments. In C. M. Reigeluth (Ed.), *Instructional design theories and models: A new paradigm of instructional theory* (pp. 215–239). Lawrence Erlbaum Associates.

Kim, Y. J., & Ifenthaler, D. (2019). Game-based assessment: The past ten years and moving forward. In D. Ifenthaler & Y. Kim (Eds.), *Game-based assessment revisited. Advances in game-based learning*. Springer.

Kluger, A. N., & DeNisi, A. (1996). Effects of feedback interventions on performance: A historical review, a meta-analysis, and a preliminary feedback intervention theory. *Psychological Bulletin, 119*, 254–284.

Merrill, M. D. (2002). A pebble-in-the-pond model for instructional design. *Performance Improvement, 41*(7), 41–46.

Merrill, M. D. (2013). *First principles of instruction: Identifying and designing effective, efficient and engaging instruction*. Pfeiffer.

Milrad, M., Spector, M., & Davidsen, P. (2003). Model facilitated learning. In S. Naidu (Ed.), *Learning and teaching with technology: Principles and practices* (pp. 11–24). Kogan Page.

Moxnes, E., & Jensen, L. (2009). Drunker than intended: Misperceptions and information treatments. *Drug and Alcohol Dependence, 105*(1-2), 63–70.

Moxnes, E., & Saysel, A. K. (2009). Misperceptions of global climate change: Information policies. *Climatic Change, 93*(1-2), 15–37.

Narciss, S. (2008). Feedback strategies for interactive learning tasks. In J. M. Spector, M. D. Merrill, J. J. G. van Merrienboer, & M. P. Driscoll (Eds.), *Handbook of research on educational communications and technology* (pp. 125–144). Lawrence Erlbaum Associates.

Narciss, S. (2013). Designing and evaluating tutoring feedback strategies for digital learning environments on the basis of the interactive tutoring feedback model. *Digital Education Review, 23*, 7–26.

Narciss, S. (2017). Conditions and effects of feedback viewed through the lens of the interactive tutoring feedback model. In D. Carless, S. M. Bridges, C. K. Y. Chan, & R. Glofcheski (Eds.), *Scaling up assessment for learning in higher education* (pp. 173–189). Springer.

Pavlov, O. V., Saeed, K., & Robinson, L. W. (2015). Improving instructional simulation with structural debriefing. *Simulation & Gaming, 46*(3-4), 383–403.

Ramaprasad, A. (1983). On the definition of feedback. *Behavioral Science, 28*, 4–13.

Reigeluth, C. M. (1999a). What is instructional design theory and how is it changing? In C. M. Reigeluth (Ed.), *Instruction design theories and models. A new paradigm of instructional theory* (pp. 5–30). Erlbaum.

Reigeluth, C. M. (1999b). The elaboration theory: Guidance for scope and sequence decisions. In C. M. Reigeluth (Ed.), *Instructional design theories and models: A new paradigm of instructional theory* (pp. 425–453). Erlbaum.

Richardson, G. P. (2014). Model teaching II: Examples for the early stages. *System Dynamics Review, 30*(4), 283–290.

Sadler, D. R. (1989). Formative Assessment and the Design of Instructional Systems. *Instructional Science, 18*(2), 119–144.

Seel, N. M., Lehmann, T., Blumschein, P., & Podolskiy, O. A. (2017). *Instructional design for learning: Theoretical foundations*. Springer.

Shute, V. J. (2008). Focus on formative feedback. *Review of Educational Research, 78*(1), 153–189.

Spector, J. M. (2000). Towards a philosophy of instruction. *Educational Technology & Society, 3*(3), 522–525.

Spector, J. M. (2009). Adventures and advances in instructional design theory and practice. In L. Moller, J. B. Huett, & D. M. Harvey (Eds.), *Learning and instructional technologies for the 21st century* (pp. 1–14). Springer.

Spector, J. M., & Anderson, T. M. (2000). *Integrated and holistic perspectives on learning, instruction and technology*. Kluwer Academic Publishers.

Sterman, J. D. (1994). Learning in and about complex systems. *System Dynamics Review, 10*(2-3), 291–330.

Sterman, J. D. (2002). System Dynamics: Systems thinking and modeling for a complex world. In *Proceedings of the ESD internal symposium*.

Tversky, A., & Kahneman, D. (1974). Judgment under uncertainty: Heuristics and biases. *Science, 185*(4157), 1124–1113.

Van der Kleij, F. M., Feskens, R. C., & Eggen, T. J. (2015). Effects of feedback in a computer-based learning environment on students' learning outcomes: A meta-analysis. *Review of Educational Research, 85*(4), 475–511.

van Merriënboer, J. J., & Kirschner, P. A. (2017). *Ten steps to complex learning: A systematic approach to four-component instructional design*. Routledge.

Wood, D., Bruner, J. S., & Ross, G. (1976). The role of tutoring in problem solving. *Journal of Child Psychology and Psychiatry, 17*(2), 89–100.

Chapter 3
Theoretical Framework

The literature discussed in Chap. 2 and a focus group discussion[1] carried out with two experts reveal that students experience multifaceted challenges when they try to understand and communicate their understanding about CDS. Hence, the primary objective of this monograph is enhancing students' learning in and about CDS. For this purpose, the authors aimed at developing a personalized and adaptive OILE that supports and facilitates the students learning. To achieve the goal of enhancing students' learning in and about CDS, the authors planned for the OILE to have the following three characteristics:

1. It presents an authentic, complex dynamic problem that the learner should address in its entirety. It then proceeds to allow learners to progress through a sequence of gradually more complex learning tasks.
2. It allows for the learner to interact with the OILE while solving the problem at hand. Upon the completion of each learning task and based on their individual performance, the OILE provides the learners with information intended to facilitate the learning process. The support fades away as learners gain expertise.
3. It tracks and collects information on students' progress and generates learning analytics that are being used to assess students' learning and to tailor the information feedback to the students.

This chapter presents the theoretical framework of the research that has been used to design the OILE with the above three features. The first section of this chapter presents a general theoretical framework for designing personalized and

[1]As part of the OILE development, a focus group discussion was conducted with two professors who teach two System Dynamics courses to master program students at the University of Bergen. The discussion was focused on three questions: What challenges students often experience when they try to understand CDS? What do students need to do to develop a comprehensive understanding of CDS? How could teachers and instructional designers support students in their learning in and about CDS?

A. T. Tadesse et al., *Adapting Interactive Learning Environments to Student Competences*, SpringerBriefs in Educational Communications and Technology,
https://doi.org/10.1007/978-3-030-88289-1_3

adaptive OILE. The next section of the chapter presents an OILE, the Mr. Wang Bicycle Repair Shop OILE, to demonstrate how the theoretical design framework of the research has been applied practically to design a personalized and adaptive OILE that support students learning in and about CDS. The last section of the chapter discusses how the questions (items) of the Mr. Wang Bicycle Repair Shop OILE and the associated feedback, the scaffolding feedback, have been designed and integrated with the OILE to support the students' learning in and about CDS.

3.1 Theoretical Framework for Designing Personalized and Adaptive OILE

A holistic instructional design (Spector & Anderson, 2000; van Merriënboer & Kirschner, 2017) has been applied in five steps to create a personalized and adaptive online interactive learning environment to support students in their study of complex, dynamic systems. The five steps include:

1. Identify instructional design models
2. Identify authentic (real-world) learning tasks
3. Identify instructional methods
4. Identify instructional techniques
5. Design interface and implement the tool

Table 3.1 summarizes the five steps of the OILE design framework and provides the main references upon which the framework has been based.

In the research, there is a strong believe that when designing learning environments that supports students learning in and about CDS, the first step should be the identification of the proper instructional design model(s). Because it is this (these) instructional design model(s) that should give 'explicitly guidance on how to better help people learn and develop' as Reigeluth (1999) underscored in his definition of instructional design theory. The instructional design model could be one or could be an assembly of instructional design models. As discussed in Chap. 2, six well-documented instructional design models have been considered in the design of the personalized and adaptive OILE; 4C/ID, First Principles of Instruction, CLE, TCI, Cognitive Apprenticeship, and Elaboration Theory.

The authors of these six instructional design models agree when and what kind of learning material should be used to help students learning in and about CDS. They all agree that an instructional designer who has chosen to use either or all of these instructional design models should first identify the nature of the learning task(s) and the task(s) should be authentic. In line with the view of the authors of these instructional design models in this research the identification of authentic learning tasks is considered as the second step of the OILE design framework. As stated above, the OILE has been planned to have three features. The first of the three features is—it presents an authentic, complex, and dynamic problem that learners

Table 3.1 A five-step design framework for the OILE, adopted from Tadesse and Davidsen (2019)

Identify instructional design models		Identify authentic learning tasks	Identify instructional methods		Identify instructional techniques	Design interface and implement the tool
Holistic instructional design	The Four Component Instructional Design model (4C/ID, van Merriënboer & Kirschner, 2017)	Causes of Oscillation: Mr. Wang Bicycle Repair Shop Case Study	Instructional scaffolding method (Wood et al., 1976; Belland, 2017)	Dynamic assessment	Multiple-choice questions and open-end questions	• Welcome page • Learning task presentation page • Supportive information provision page • Navigation buttons (OECD, 2013; Alessi & Trollip, 2001)
	First Principles of Instruction (Merrill, 2002, 2013)			Providing just the right amount of support	• Story telling: linking previous and current learning tasks • Repeated trial: giving multiple opportunities to try a question • Providing feedback and feed-forward • Item branching: branching to either a simpler or more complex concept (Jonassen, 1999; Reigeluth et al., 2017; van Merriënboer & Kirschner, 2017; Merrill, 2013)	
	Constructivist Learning Environments (CLE, Jonassen, 1999)					
	Task Centered Instruction (TCI, Francom & Gardner, 2014; Francom, 2017)					
	Cognitive Apprenticeship (Collins et al., 1989, 1991)			Intersubjectivity	• Providing part-task practice (van Merrienboer & Kirschner, 2017) • Providing summary	
	Elaboration Theory (Reigeluth, 1999).					

should address in its entirety. In the research, the Mr. Wang Bicycle Repair Shop case study, which is an authentic, complex, and dynamic problem has been considered for learners to address it in its entirety to gain all constitute skill they should acquire. Such skills comprise problem conceptualization, model building, model analysis, and policy design. The case study has been designed to teach System Dynamics master program students at the University of Bergen about the cause of oscillation.

Following the identification of the learning material, the six instructional design models recommend designers to consider appropriate instructional methods to support and facilitate the students' learning and so does the third step of the design framework of this research. In this research, the OILE has been planned to have a second feature that allows students to interact with it while the students are solving the complex and dynamic problem. Moreover, as this second feature, it was planned that, upon the completion of each learning task, the OILE will provide the learners with supportive information based on their individual performance. This support fades away as the learners gain expertise. To provide such support and configure such a feature with the OILE, the instructional scaffolding method (Wood et al., 1976; Belland, 2017) has been considered in the research.

In the next stage of the instructional design, the six instructional design models recommend the chosen instructional method(s) to be broken into simpler and more detailed component methods to provide explicit guidance on how to support and facilitate the students learning. The theoretical framework of this research acknowledges this step of the design and suggests identifying instructional techniques as the fourth step of the OILE design. The instructional scaffolding method has three component elements: dynamic assessment, provision of just the right amount of support, and intersubjectivity (Belland, 2017). These three elements of the instructional scaffolding method were used in the research to design the instructional techniques that guided the development of the OILE. One of the instructional techniques integrated with the OILE is scaffolding feedback. Scaffolding feedback is an educational feedback that provides information to the students through the OILE regarding a gap in performance between the students' actual performance and a previously set desired performance to alter the gap. A detailed account of the scaffolding feedback and its design is presented in the last section of this chapter.

The third desired feature for the OILE to possess is the ability to track and collect information on the learners' progress and generate learning analytics, which would be used to assess the students' learning. To implement this feature in practice, the identified instructional techniques need to be manifested in the form of an educational tool, the OILE. And the OILE need to be made available to students with information storing features. Consequently, in this research, we suggest the last and fifth step of the OILE theoretical framework to be designing the OILE's interface and practically implementing the educational tool in a digital platform.

In the research, the Mr. Wang Bicycle Repair Shop OILE has been designed on the interface of a computer modeling software called STELLA Architect, which offers data collecting features (https://www.iseesystems.com/store/products/stella-architect.aspx). The data collection feature of the software was created after the authors of this research proposed the inclusion of the feature into the modeling software and through the good will of the isee Systems company (https://www.iseesystems.com/).

3.2 The Mr. Wang Bicycle Repair Shop OILE

The Mr. Wang Bicycle Repair Shop OILE (abbreviated to the Mr. Wang OILE) is an on-line (web-based) learning environment (OILE) built using the Stella Architect software. The Mr. Wang OILE has been designed around the Mr. Wang Bicycle Repair Shop case, which is an authentic, complex, and dynamic problem. The case study concerns a reputable bicycle repair shop in Shanghai that repairs and delivers your bicycle in top shape after one day. The case study is designed to teach master students in the System Dynamics program at the University of Bergen about the causes of oscillation.

Oscillation is one of the fundamental modes of behavior produced by feedback systems. It occurs virtually in all business areas such as in commodity markets, labor supply chains, manufacturing supply chains, and real estate markets. Using this case study, the students investigate the causes of oscillation experienced by the Bicycle Repair shop. The case study has two parts. The first part focuses on problem identification and analysis, whereas the second part is on policy formulation and analysis. The first part of the case study was used in the design of the OILE, whereas, the second part was offered in a traditional paper and pencil format.

The Mr. Wang OILE has the three characteristics described at the first section of this chapter:

1. It presents an authentic, complex dynamic problem that the learner should address in its entirety. It then proceeds to allow learners to progress through a sequence of gradually more complex learning tasks.
2. It allows for the learner to interact with the OILE while solving the problem at hand. Upon the completion of each learning task and based on their individual performance, the OILE provides the learners with information intended to facilitate the learning process. The support fades away as learners gain expertise.
3. It tracks and collects information on students' progress and generates learning analytics that are being used to assess students' learning and to tailor the information feedback to the students.

When students first enter the Mr. Wang OILE, they receive a quick introduction about the objective of the Mr. Wang Bicycle Repair Shop case study (Fig. 3.1). A description of the buttons, which help them navigate through the learning environment, follows in the next page. In the subsequent pages, the students are presented with Mr. Wang's description of his Bicycle Repair Shop and the dynamic problem he has been experiencing in his Repair Shop.

Following Mr. Wang's introduction of the problem, students procced to the different tasks of the case study. The first part of the case study, which has been used

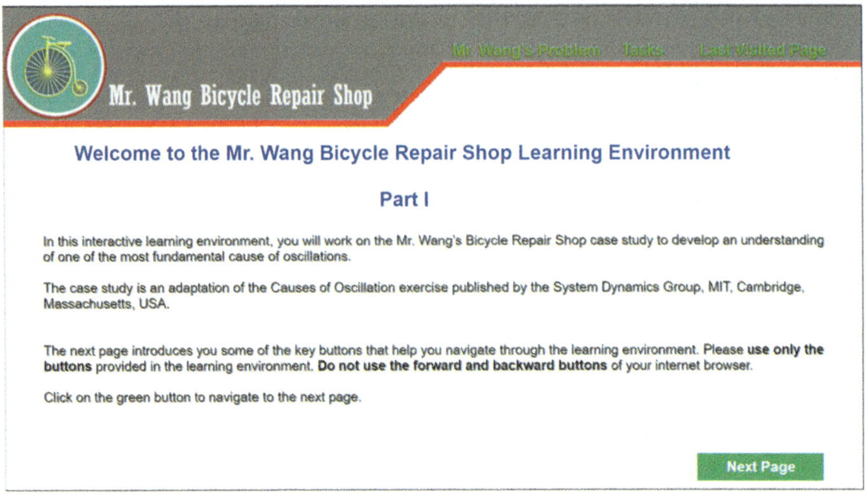

Fig. 3.1 Welcome page for the Mr. Wang OILE

in the development of the OILE, is divided into five tasks. A task is a subset of items (questions/challenges) with specific objective(s), which students should be able to achieve upon completion of that task. Richardson (2014) calls these kind of tasks "canned modeling exercises", where a "student opens the 'can' and out tumbles all the ingredients needed to formulate [a computer] model [that represent the problem under study]" (p. 295).

The five different tasks were categorized under three parts of the Mr. Wang OILE. The first part of the OILE consists of Task 1 and Task 2. The second part has Task 3 and Task 4 and the third part consists of Task 5. The first task focuses on problem identification and definition. In this task, the learners are required to identify the problem of the Repair shop. Tasks 2–5 concentrate on hypothesis formulation and analysis of that hypothesis. In these tasks, the students are required to formulate a hypothesis about the underlying causal structure of the problem. They then proceed to analyze the relationship between that structure and the consequent dynamic behavior by building a computer model. The students carry out this task in a reiterative process until they arrive at a structure that best explains the identified problem. The complexity of the underlying causal structure and its analysis increase as the students proceed from Task 2 to Task 5.

There are 116 items (questions) that all the students are supposed to complete, and 16 additional items are offered to those who might need extra support. The 116 items were divided across the three parts of the Mr. Wang OILE. The first 25 items were organized under the first part of the Mr. Wang OILE, the next 40 items belong to the second part of the OILE and the last 51 items were offered under the third part of the OILE.

3.3 Item and Scaffolding Feedback Design for the Mr. Wang OILE

The six instructional design models that influenced the development of the Mr. Wang OILE recommend that the instructional method considered in the design, the instructional scaffolding method, be broken into simpler and more detailed component methods to provide explicit guidance on how to support and facilitate the students learning. The instructional scaffolding method has three component elements that provided explicit guidance in the design of the Mr. Wang OILE; dynamic assessment, provision of the right-support, and intersubjectivity (Belland, 2017). The first two components, dynamic assessment and provision of the right-support, have been used in the design of items and scaffolding feedback of the Mr. Wang OILE, respectively. Whereas, the third component, intersubjectivity, has been considered during the design of both the items and the scaffolding feedback of the OILE.

3.3.1 Item Design for the Mr. Wang OILE

The dynamic assessment in the Mr. Wang OILE has been done using multiple-choice questions (MCQ) and open-ended questions (OEQ). The format of the MCQ has been consistent throughout the OILE with four alternatives, except in two specific questions that have only two alternatives. In the Mr. Wang OILE, there is only one correct choice per question, and learners can only choose one answer at a time. The items in the OEQ format ask students to predict over time changes in a variable(s) of importance. The students provide their response through the OILE using graph drawing tools that have been integrated with the Mr. Wang OILE.

The learning material of the Mr. Wang OILE has been designed in a way that, at each stage of the learning activity, an item is posed for the students to solve. The learners work on the item and provide their response either by choosing one of the MCQ alternatives or by drawing behavior graphs. The items range from identifying a vivid problem to hypothesizing a causal structure responsible for that problem and analyzing the relationship between suggested structure and the consequent dynamic behavior by building computer models. During the model-building phase, learners are required to work with a modeling software pre-installed on their local computers. In this stage, the learners are often required to switch between the OILE and the modeling software, so that they engage in hands-on activities.

The Mr. Wang OILE items have been arranged in sequences called 'learning paths'. A learning path is a sequence of items that learners pass through, while working on the complex and dynamic problem on their own pace and time. Each learner has her/his own unique learning path. In general, there are linear and branching sequence questions in the learning path of a learner. Linear sequence questions are those where a learner moves to the next question after finishing the previous question without any precondition. Branching sequence questions are those where the next question depends on the performance of the learner when responding to the previous question. The branching technique is discussed in the next subsection.

Intersubjectivity has been maintained using the items of the Mr. Wang OILE by allowing students to pass through iterative steps at each stage of the learning material. Every time students are required to conceptualize a problem, for example, they will be asked first to identify a variable that represents the symptom of the problem (stock/accumulator). Then they will be asked to identify the variables that cause the stock/accumulator to change (flows) and finally variables that influence the flow rates to change (auxiliary variables or parameters). van Merriënboer and Kirschner (2017) classify such skills as 'recurrent constitute skills'. These 'recurrent' skills can be acquired either by following certain procedures and/or rules or by "continually practicing them in order to automate those constitute skills" (van Merriënboer & Kirschner, 2017, p. 97). However, skills such as identifying a stock or a flow variable from a problem description can be achieved by building schema of those variables (Jonassen, 2000). van Merriënboer and Kirschner (2017) classify these skills as 'non-recurrent constitute skills'. The techniques used in the design of the items of the Mr. Wang OILE were aimed at strengthening the construction of both 'recurrent' and 'non-recurrent' constitute skills, thereby establishing intersubjectivity between the scaffoldee and the scaffolder.

When the students are required to analyze the behavior of a model output, they will be asked to chop overtime changes of the model behavior based on the need to identify monotonic behavior developments, so as to explain each monotonic development component by referring back to the structure of the model. In doing so, the students can practice and strengthen their analytical skills. van Merriënboer and Kirschner (2017) call this 'part–task practice'. The part-task practices were designed to help the students develop particular sub-skills and acquire automatic performance.

3.3.2 Scaffolding Feedback Design for the Mr. Wang OILE

To provide the right-support and also establish intersubjectivity between the scaffoldee and the scaffolder using the Mr. Wang OILE, the scaffolding feedback has been designed and communicated to the students by way of the OILE.

The design of the scaffolding feedback followed three feedback design strategies: (1) Identify focus areas of competencies, which the feedback helps to improve (feedback levels). (2) Identify the kind of feedback that can be communicated to improve the identified focus areas of competencies (type of feedbacks). (3) Determine the time at which the identified feedback type can be communicated to improve the selected focus areas of competencies (feedback timing).

3.3.2.1 Identifying the Feedback Levels

Educational feedback literature shows that students' competency level can be improved by targeting the feedback at four different feedback levels (focus areas of competency): at item level [how well items are understood/performed], at item process level [effective item processing strategies used], at self-regulation level [self-monitoring, directing, and regulating action], and at self level [evaluating personal aspect—praise/reward, motivation] (Hattie & Timperley, 2007).

During the focus group discussion with the two professors, who teach two different System Dynamics courses to the master program students at the University of Bergen, four focus areas of improvement were identified.

1. Help students develop the skill of recognizing problematic behavioral modes from problem descriptions.
2. Help students develop the skill of model building as part of problem conceptualization.
3. Help students develop the skill of identifying changes in behavioral modes
4. Help students develop the skill of analyzing behavior inferring from its underlying structure

To help students develop these competencies, support was provided to students using the Mr. Wang OILE at three of the four feedback levels: At item level, at item process level, and at self-regulation level. The support was provided to students regarding certain aspects of the items of the Mr. Wang case study and on how to complete those items more effectively. At the same time, the support was aimed at empowering the students to be able to evaluate their own performance and identify the correct performance from the one that was not, in the process enabling them develop self-regulation.

3.3.2.2 Identifying the Feedback Types

To help students improve the competencies identified during the focus group discussion, different feedback types were considered in the design of the scaffolding feedback. These feedback types include Knowledge of result/response (KR), Knowledge of the correct response (KCR), and Elaborated feedback (EF). These feedback types were utilized to different degrees in the design of the scaffolding feedback and were synergized to form the scaffolding feedback.

The scaffolding feedback of the Mr. Wang OILE has three elements: Storytelling (feed-up), feed-back, and feed-forward. These three elements of the scaffolding feedback are consistent with Sadler (1989) and Hattie and Timperley (2007) suggestions for effective feedback. They suggest that an effective feedback should comprise elements that answer at least three questions: Feed-up (goal)—Where am I going? Feed-back—How am I going? and Feed-forward—Where to next? In the research, the identified feedback types were integrated with the three elements of the scaffolding feedback to improve the students' competencies identified during the focus group discussion.

The storytelling (feed-up) element has been used to present the content of the learning material and to contextualize the students' learning. This instructional technique was applied to link what students already know with the new information. It was also used to provide important information to learners that might help them solve the learning tasks. In general, the story telling was used to inform learners where they are going by informing them about the objective of the question and what they are about to address. In other words, the feed-up has been targeted at item level—providing important and necessary information about the item to be solved. This technique has been used in almost all of the six instructional design models and other instructional design theories that influenced the design framework of this research, but under different names. Jonassen (1999) and Reigeluth et al. (2017) call it adjusting scaffolding, while Merrill (2013) and Francom and Gardner (2014) name this technique activation of prior knowledge. From Robert Gagné (1985) nine events of instruction,[2] this technique comprises the first three: Gain attention of the students, inform students of the objectives, and stimulate recall of prior learning.

In the Mr. Wang OILE, feed-back—the second element of scaffolding feedback, was communicated to the students in different forms. The first was in the form of suggested answer to the open-ended question so that the students can compare their response with the suggested answer. This feedback type is called Knowledge of the correct response—KCR (Mory, 2004; Narciss, 2008; Shute, 2008). The second was in the form of repeated trials. The repeated trial option gives students the opportunity to try a question multiple times with varying level of support. The feedback literature calls this type of feedback elaborated feedback—EF (Mory, 2004; Shute, 2008; Narciss, 2008, 2013; Van der Kleij et al., 2015).

The Mr. Wang OILE offers the students three opportunities to try and correctly answer a multiple-choice question that has four alternatives. A student who fails to respond correctly to a question twice receives more support than those who fail only once.

Repeated wrong choices by students' were used as a proxy for identifying possible misconceptions, which teachers could address during face-to-face instructions. Also, such repeated wrong choices can help teachers and instructional designers to identify learning tasks that need revision to improve the quality of the OILE. Particularly, if large number of students missed a particular question, it might be important to check whether that question was properly presented. From the

[2] Robert Gagné (1985) proposed a series of events of instruction that can be offered to students as external support to help them develop their internal knowledge about materials the students are studying. Gagné (1985) calls these external supports as 'nine even of instruction' that include: Gaining attention, Informing learners of the objective, Stimulating recall of prior learning, Presenting the stimulus, Providing learning guidance, Eliciting performance, Assessing performance, and Enhancing retention and transfer.

students' perspective, repeated wrong answers might help them recognize their performance level and their progress in the learning tasks.

Unlike in the case of the OEQ, where all the students receive identical feedback, in the MCQ the students can receive different feedback based on their individual performance. Students who select the correct answer to a question receive a scaffolding feedback, which inform them why that specific alternative is correct and why the other alternatives are wrong. This feedback type is called Knowledge of concept feedback—part of the EF (Narciss, 2013). Such feedback was designed with two specific objectives: (1) To make sure that the students know the correct reason why their responses are correct, thereby strengthening intersubjectivity between the scaffoldee and the scaffolder. (2) To prevent the impact of "guessing" on subsequent tasks, thereby serving as a feed-forward—showing the direction for where next.

If the students' response are wrong and are not their third trial, they can receive either a corrective feedback or an ordinary feedback with item branching.

A corrective feedback is a response-contingent feedback (part of the EF, Narciss, 2013) that explicitly shows the reason why the students' answers are wrong. Whereas an ordinary feedback with item branching is a combination of two feedback. The first part is Knowledge of response (KR) feedback that simply tells the students their reply is incorrect. The second part is Knowledge about meta-cognition (part of the EF, Narciss, 2013). Unlike the corrective feedback, the students do not receive information about why their answers are wrong. Rather, the students are asked to branch to other questions that are easier than the previous question, but under the same conceptual framework, so that they can figure out on their own why their previous responses were wrong.

The option for providing either corrective feedback or ordinary feedback with item branching dependents on individual students' learning paths and the stages at which the students are in the learning tasks.

Students receive corrective feedback while they are in the early stages of problem identification of the Mr. Wang case study. Students' continued engagement in the learning environment can be affected by their early perception regarding what they are going to do (Jonassen, 1999). Hence, the learning tasks and the support provided during the early stages of the learning process should help the learners understand the problem statements clearly.

As the students advance through the Mr. Wang learning material, corrective feedback gradually fades away and are replaced by ordinary feedback with item branching. The students can branch up to three levels. Those students who fail to respond correctly at the lowest level receive corrective feedback. Students, who are successful in responding to the lower level questions, move up to higher levels and work again the questions they failed to respond correctly to. In doing so, the students move up and down the ladder of the Mr. Wang OILE. Students who respond correctly to the questions at the top level progress to more complex tasks. As the learners progress through the learning material and gain more expertise, the item branching reduces from three levels to two levels and finally to one.

3.3.2.3 Identifying the Feedback Timing

In this research, the scaffolding feedback has been designed to be offered through the Mr. Wang OILE and there was no delay in the provision of the feedback. All the feed-back and feed-forward elements of the scaffolding feedback were communicated to the students immediately after they responded to an item and the feed-up was presented before they started working on the item. Hence, in this research the option of providing a delayed feedback (Kulhavy & Anderson, 1972; Phye & Andre, 1989; Shute, 2008) was not considered.

References

Alessi, S. M., & Trollip, S. R. (2001). *Multimedia for learning: Methods and development* (3rd ed.). Allyn & Bacon.

Belland, B. (2017). *Instructional scaffolding in STEM Education: Strategies and efficacy evidence*. Springer Open.

Collins, A. M., Brown, J. S., & Holum, A. (1991). Cognitive apprenticeship: Making thinking visible. *American Educator, 15*(3), 6–11.

Collins, A. M., Brown, J. S., & Newman, S. E. (1989). Cognitive apprenticeship: Teaching the crafts of reading, writing, and mathematics. In L. B. Resnick (Ed.), *Knowing, learning, and instruction: Essays in honor of Robert Glaser* (pp. 453–494). Mahwah, NJ: Lawrence Erlbaum Associates.

Francom, G. M., & Gardner, J. (2014). What is task-centered learning? *TechTrends, 58*(5), 27–35.

Francom, G. M. (2017). Principles for task-centered instruction. In C. M. Reigeluth, B. J. Beatty & R. D. Myers (Ed.), *Instructional design theories and models: The learner-centered paradigm of education* (Vol 4, pp. 65-91). Taylor & Francis.

Gagné, R. M. (1985). *The conditions of learning and theory of instruction* (4th ed.). Holt, Rinehart and Winston.

Hattie, J., & Timperley, H. (2007). The power of feedback. *Review of Educational Research, 77*, 81–112.

Jonassen, D. H. (1999). Designing constructivist learning environments. In C. M. Reigeluth (Ed.), *Instructional design theories and models: A new paradigm of instructional theory* (pp. 215–239). Lawrence Erlbaum Associates.

Jonassen, D. H. (2000). Toward a design theory of problem solving. *Educational Technology Research and Development, 48*(4), 63–85.

Kulhavy, R. W., & Anderson, R. C. (1972). Delay-retention effect with multiple-choice tests. *Journal of Educational Psychology, 63*(5), 505–512.

Merrill, M. D. (2002). A pebble-in-the-pond model for instructional design. *Performance Improvement, 41*(7), 41–46.

Merrill, M. D. (2013). *First principles of instruction: Identifying and designing effective, efficient and engaging instruction*. Pfeiffer.

Mory, E. H. (2004). Feedback research revisited. In D. Jonassen (Ed.), *Handbook of research on educational communications and technology* (pp. 745–783). Lawrence Erlbaum.

Narciss, S. (2008). Feedback strategies for interactive learning tasks. In J. M. Spector, M. D. Merrill, J. J. G. van Merrienboer, & M. P. Driscoll (Eds.), *Handbook of research on educational communications and technology* (pp. 125–144). Lawrence Erlbaum Associates.

Narciss, S. (2013). Designing and evaluating tutoring feedback strategies for digital learning environments on the basis of the interactive tutoring feedback model. *Digital Education Review, 23*, 7–26.

OECD. (2013). PISA 2012 *Assessment and analytical framework: Mathematics, reading, science, problem solving and financial literacy*. Accessed on 28 May 2016 from http://www.oecd.org/pisa/pisaproducts/PISA%202012%20framework%20e-book_final.pdf

Phye, G. D., & Andre, T. (1989). Delayed retention effect: Attention, perseveration, or both? *Contemporary Educational Psychology, 14*(2), 173–185.

Reigeluth, C. M. (1999). What is instructional design theory and how is it changing? In C. M. Reigeluth (Ed.), *Instruction design theories and models. A new paradigm of instructional theory* (pp. 425–453). Erlbaum.

Reigeluth, C. M., Myers, R. D., & Lee, D. (2017). The learner-centered paradigm of education. In C. M. Reigeluth, B. J. Beatty, & R. D. Myers (Eds.), *Instructional design theories and models: The learner-centered paradigm of education* (Vol. 4, pp. 1–32). Taylor & Francis.

Richardson, G. P. (2014). Model teaching III: Examples for the later stages. *System Dynamics Review, 30*(4), 291–299.

Sadler, D. R. (1989). Formative assessment and the design of instructional systems. *Instructional Science, 18*(2), 119–144.

Shute, V. J. (2008). Focus on formative feedback. *Review of Educational Research, 78*(1), 153–189.

Spector, J. M., & Anderson, T. M. (2000). *Integrated and holistic perspectives on learning, instruction and technology*. Kluwer Academic Publishers.

Tadesse, A. T., & Davidsen, P. I. (2019). Framework to support personalized learning in complex systems. *Journal of Applied Research in Higher Education, 12*(1), 57–85.

Van der Kleij, F. M., Feskens, R. C., & Eggen, T. J. (2015). Effects of feedback in a computer-based learning environment on students' learning outcomes: A meta-analysis. *Review of Educational Research, 85*(4), 475–511.

van Merriënboer, J. J., & Kirschner, P. A. (2017). *Ten steps to complex learning: A systematic approach to four-component instructional design*. Routledge.

Wood, D., Bruner, J. S., & Ross, G. (1976). The role of tutoring in problem solving. *Journal of Child Psychology and Psychiatry, 17*(2), 89–100.

Chapter 4
Assessing the Design Framework

Numerous studies document that people have difficulties comprehending complex, dynamic systems (CDS) and communicating their understanding about such systems. Efforts were made to support learning in and about CDS. However, there are still significant gaps in our understanding of how to support and improve cognitive and communicative capabilities in and about CDS. This research aims at extending the effort of supporting students' in this regard by designing personalized and adaptive online interactive learning environment (OILE). Consequently, the overarching research question addressed was; "How may we design personalized and adaptive OILEs that effectively enhance students' learning in and about CDS?" This research question was investigated in three different studies, where the findings of one study inform the scope, goal, and design of the subsequent study. This chapter first presents the main research method documented in this monograph. It then presents the sampling strategies and the samples used to assess the OILE. In the subsequent subsections, the data collection methods and the data collection tools used in the research are being presented. Finally, there is a summary on the major findings of each of the three studies and the main questions investigated in each study.

4.1 Research Method

The main research method applied in this work is mixed methods research. Design based research is conducted using a set of research methods. Of the set of research methods applied, the most common one is the mixed methods research (Anderson & Shattuck, 2012; McKenney & Reeves, 2013). Johnson et al. (2007) define mixed method research as:

> [A] type of research in which a researcher or team of researchers combines elements of qualitative and quantitative research approaches (e.g., use of qualitative and quantitative

A. T. Tadesse et al., *Adapting Interactive Learning Environments to Student Competences*, SpringerBriefs in Educational Communications and Technology,
https://doi.org/10.1007/978-3-030-88289-1_4

viewpoints, data collection, analysis, inference techniques) for the broad purposes of breadth and depth of understanding and corroboration. (p. 123)

In a continuum of qualitative-quantitative research, the mixed methods research can be pure mixed, with equal proportion of qualitative and quantitative methods, qualitative dominant or quantitative dominant (Johnson et al., 2007). In this research, the quantitative dominant mixed methods research has been applied to conduct a survey study and two impact studies, a quasi-experiment and a single-subject experiment (sometimes referred as single-subject design, Sheskin, 2003).

A survey study is a kind of research conducted to provide "a quantitative or numeric description of trends, attitudes, or opinions of a population by studying a sample of that population" (Creswell & Creswell, 2018, p. 336). A quasi-experiment is a type of experimental study that uses a non-randomized selection and/or assignments of samples to study groups, whereas a single-subject experiment is an experimental study where a single individual or group of individuals are studied over time (Sheskin, 2003; Creswell & Creswell, 2018).

The survey study was conducted to investigate the students' attitude *towards* and experience *with* the Mr. Wang OILE using two questionnaires. The single-subject experimental study was conducted to assess changes in the students' performance over time by using their process log. The quasi-experimental study was conducted to investigate gain in treatment groups' performance compared to their comparable control groups using posttests and transfer skill exercises.

4.2 Sampling and Study Participants

Sampling, in a broader sense, is a technique of selecting participants or subjects of a study from a population to be studied (Fowler, 2014; Babbie, 2015). There are various types of sampling techniques that allow a researcher to select participants that would best describe the population under study. The ideal sampling technique is random sampling where each individual in the population has an equal probability of being selected. For a number of reasons, however, researchers tend to use a nonrandomized purposive sampling technique, where participants are chosen based on their convenience and availability (Creswell & Creswell, 2018).

In this research, the purposive sampling technique was applied to select three cohorts of System Dynamics master program students at the University of Bergen. Eighty-four students were involved in the research over a three-year period, from 2016 to 2018. For the quasi-experimental study, the first group of students (N = 27) from the academic year 2016 was used as Control group, whereas the second group of students (N = 33) from the academic year 2017 and the third group of students (N = 24) from the year 2018 were used as Experimental groups, Experimental1 and Experimental2, respectively. For the survey study and the single subject experimental study, 33 students from the 2017 and 24 students from the 2018 academic years were used as samples.

In the academic setting where this research was conducted, it was not possible to give some students an innovative new instructional tool, the Mr. Wang OILE, while denying others in the same program that same tool. This is due, in part, to avoid a possible impact of the new tool on the students' grade. Therefore, each student cohort was used as an individual Experimental or Control group. To address possible impacts of confounding variables that might come from non-randomized assignment of students to each condition, pre-assessment information was analyzed. The pre-assessment includes students' demographic information and students' performance on four different assignments, which the students completed prior to the interventions. A detailed account for the pre-assessment tools is provided in the data collection section of this chapter.

4.3 Data Collection

In the research underlying this monograph, five different data collection methods were used; pre-assessment tools, questionnaires, process log, posttest, and transfer skill exercises. These data collection methods are described in the subsections below.

4.3.1 Pre-assessment Tools

Two data sources, students' background information and their performance on four different assignments of an introductory System Dynamics course, were used to assess the equivalence of the study participants before they were assigned to the treatment groups of the quasi-experimental studies.

Students' background information, such as age, gender, and program type, was used as the first pre-assessment tool. Soon after students are admitted to the master program in System Dynamics at the University of Bergen, they take a 5-week System Dynamics introductory course (10 ECTS credit course), a mandatory course in the master program. During this period, the students are introduced to the theory, methods, and tools of System Dynamics. They are introduced to the basics of dynamics systems (a stock with in- and out-flows, local feedback from stock to own flows, nonlinearities, and major loops with delays) and the use of causal loop diagrams, table functions, and equations to represent and illustrate cause-and-effect relationships. During this period, the students complete four identical assignments. The assignments are part of their mandatory course work, and to sit for the final exam of the introductory course, the students are required to complete all the four assignments. These four assignments were used as the second pre-assessment tools of the research. Each assignment was scored from a maximum of 3 points. Hence, a student could attain a total maximum points of 12 or a minimum points of zero from the four assignments.

4.3.2 *Questionnaires*

Two questionnaires (survey research method, Fowler, 2014), based on prior research (Taylor-Powell & Renner, 2009; Maor & Fraser, 2005; Berkeley Center for Teaching and Learning, n.d.), were used to assess the affective aspects of students learning after their experience with the Mr. Wang OILE.

The two questionnaires, Q1 and Q2, were designed to assess the students' attitude *towards* and their experience *with* the Mr. Wang OILE. A total of 38 questions were administered by way of the two questionnaires, 17 questions in the first questionnaire and 21 questions in the second. Of the 38 questions, 33 of them are five-level Likert scale questions that range from strongly agree to strongly disagree. The remaining 5 questions, 2 from Q1 and 3 from Q2, are open-ended questions.

The Mr. Wang case study has two parts: Part I—problem identification, model building, and analysis and Part 2—policy analysis. The first part of the case study has been used in the design of the Mr. Wang OILE (see Chap. 3). The Mr. Wang OILE has three parts with five tasks. The first part of the OILE has the first two tasks, Task 1 and 2. The second part has Task 3 and 4 and the third one has Task 5. The first questionnaire was administered as soon as the students had completed the first part of the OILE, Tasks 1 and 2, whereas the second questionnaire was administered after they had completed the remaining three tasks, Tasks 3, 4 and 5.

4.3.3 *Process Log*

To access how well the students performed in the subsequent tasks of the case study while supported by the Mr. Wang OILE, their process log information was collected. Once a student logs into the Mr. Wang OILE, a special tracker built onto the Stella Architect software tracks the student's process log information. The tracker records information such as name, performance in the case study, the pages the student has navigated, and the amount of time the student has spent on each page. It also records what kind of support the student has received. The process log information was collected in the form of comma separated values (csv) files and time series graphs. On the basis of these pieces of information, students' learning paths were drawn.

A learning path is a sequence of questions that a student pass through, while working on the complex and dynamic problem in their own pace and time. There were 116 questions that all the students were supposed to complete in the OILE and 16 additional questions to those who might need extra support. Each learner traversed their own unique learning path, conditioned by their performance.

4.3.4 Posttest

As described above, the Mr. Wang case study has two parts. Part I of the case study was used to design the Mr. Wang OILE, whereas part II of the OILE was used as a posttest to measure differences in problem solving skills, if there were any, between those who used the OILE and those who did not.

In its original pencil and paper format, Control group students were introduced to the first part of the Mr. Wang case study by a professor. The students submitted the first part of the case study after working for a week. In case they encountered challenges while working on the case study, they consulted the teaching assistants. After submission, the professor reviewed the first part of the case in interaction with the students. Following the review, students worked on the second part of the case study for half a week. Again, after submission, the professor reviewed the second part of the case in interaction with the students. The Control group completed the case study during October 2016.

During October 2017 and October 2018, the Experimental groups, Experimental1 and Experimental2, respectively, used the Mr. Wang OILE to carry out the first part of the case study. In its new format, the professor introduced concepts relevant to the case study whereupon the students, in the course of the following week, worked entirely using the OILE in their own time and at their own pace. After submission, the professor reviewed the first part of the case in interaction with the students. Following the review, the Experimental groups worked on the second part of the case study in the same way as the Control group did (using pencil and paper format).

In part II of the case study, the students were required to introduce a policy that would solve the Mr. Wang problem that they identified and analyzed in part I of the case study. The students had half a week (4 days) to complete part II of the case study. The students' performance was scored out of 10 points. Hence, the maximum points a student could score from the posttest was ten and the minimum was zero.

4.3.5 Transferable Skill Exercise

During October 2018, the Experimental2 group students, after they completed the Mr. Wang OILE and before they started part II of the Mr. Wang case study, were asked to work on a new case study, the Mrs. Lee case study, using the pencil and paper format. The Mrs. Lee case study, which is conceptually and mathematically similar to the Mr. Wang case study, was administered to assess whether the students was able to transfer the skills they acquired while working on the Mr. Wang case study with the support of the OILE to a similar problem context.

Both the Mr. Wang and the Mrs. Lee case studies address inventory management issues. However, the former deals with a situation in a bicycle repair shop and the later deals with a case in a bicycle manufacturing shop. For detailed accounts of the case studies, please see Appendices 1 and 2.

Similar to the Mr. Wang case study, the Mrs. Lee case study has two parts to it and was designed to teach students about the causes of oscillation. After completing the first part of the Mrs. Lee case study, the Experimental2 students completed the second part of the case study, which is the same as part two of the Mr. Wang case study. Since Part 2 of the Mr. Wang and the Mrs. Lee case studies are the same, the students did only one of the Part 2 of the case studies. They did part two of the case studies in pencil and paper format, in the same way as the Control and the Experimental1 group students did.

4.4 Results

The section below presents results from three different studies. In the first study, literature of relevance to the study was explored and a focus group discussion was carried out to understand the extent of the problem associated with the teaching and learning of CDS. Also, a theoretical design framework was developed for a personalized and adaptive OILE. The design framework was used to develop the Mr. Wang OILE, which served as a basis for conducting experimental impact studies in the second and the third studies. In this first study, the students' affective domains were assessed. The second study investigated the impact of using the Mr. Wang OILE on the students' problem-solving skills—an impact that was being measured using the posttest. The third study investigated the impact of using the OILE on the students' transferable skills—measured using the transferable skill exercises. The subsections below provide a summary of the major findings of each of the three studies and the main questions investigated in each study.

4.4.1 Study I: Survey Study

The main research question investigated under this first study was: How may one design online interactive learning environments to support individual students learning in and about complex, dynamic systems?

The first study came to four major findings and these findings are summarized as follows: The first main finding was that students' difficulties in comprehending CDS and communicating their understanding about such systems arise from limitations in three different types of capabilities: (1) The cognitive capability to comprehend structural complexity. (2) The skills required to infer the dynamic behavior of a system from its underlying structures. (3) The effectiveness of methods, techniques, and tools that are available to us in our analysis of such systems (Davidsen, 1996; Spector & Anderson, 2000; Jonassen, 2000; Ifenthaler & Eseryel, 2013; van Merriënboer & Kirschner, 2017). Following this finding, efforts were made to identify ways to improve the effectiveness of methods, techniques, and tools that facilitate the teaching and learning of CDS. For this purpose, personalized and adaptive online interactive learning environments were identified and theoretical design principles were formulated based on existing literature and learning theories.

Hence, the second main finding of Study I was the formulation of a five-step theoretical design framework that includes:

1. Identify instructional design models
2. Identify authentic (real world) learning tasks
3. Identify instructional methods
4. Identify instructional techniques
5. Design interface and implement the tool

This design framework was used to design the Mr. Wang OILE, the third main finding of Study I, that has the following three features:

A. It presents an authentic, complex dynamic problem that the learner should address in its entirety. It then proceeds to allow learners to progress through a sequence of gradually more complex learning tasks.
B. It allows for the learner to interact with the OILE while solving the problem at hand. Upon the completion of each learning task, and based on their individual performance, the OILE provides the learners with information intended to facilitate the learning process. The support fades away as learners gain expertise.
C. It tracks and collects information on students' progress and generates learning analytics that are being used to assess students' learning and to tailor the information feedback to the students.

The first two features aim at enhancing the students' cognitive and communicative capabilities, whereas the third one aims at measuring the development of the students' capabilities.

The Mr. Wang OILE is an online (web-based) learning environment built using the Stella Architect software (https://www.iseesystems.com/store/products/stellaarchitect.aspx). The students used the Mr. Wang OILE during the academic years of 2017 and 2018. The students were asked to provide their attitude *towards* and their experience *with* the Mr. Wang OILE using two questionnaires, Q1 and Q2.

Hence, the fourth main finding of Study I was related to the students' response to the two questionnaires (see Table 4.8 in Appendix 3). Analysis of both Q1 and Q2 show that the students firmly believe they have been through an effective learning experience while working within the Mr. Wang OILE and they recommended other students to use the OILE in their study of the causes of oscillations.

The result of the Wilcoxon Signed-Ranks test (Table 4.1), which was conducted in order to evaluate whether there were significant differences in the students' satisfaction in Q1 and Q2, reveals that the students' attitude *towards* the feedback provided by the OILE, and their perception of their own learning, were significantly higher in the Q2, compared to the Q1. However, for the other questions, the Wilcoxon Signed-Ranks test indicates that there were no statistically significant differences in the students' satisfaction level between the two questionnaires.

Table 4.1 Results of the Wilcoxon signed-Ranks test, adopted from Tadesse and Davidsen (2019)

		N	Mean rank	Sum of ranks	Z	Asymp. Sig. (2-tailed)
Q2q1—Q1q1 (Experience with the user interface of the OILE)	Negative ranks	2[a]	2.50	5.00	.000[d]	1.000
	Positive ranks	2[b]	2.50	5.00		
	Ties	49[c]				
Q2q2—Q1q2 (Attitude towards the learning task)	Negative ranks	1[a]	3.50	3.50	−1.633[e]	.102
	Positive ranks	5[b]	3.50	17.50		
	Ties	47[c]				
Q2q3—Q1q3 (Attitude towards the feedback offered)	Negative ranks	1[a]	6.00	6.00	−2.714[e]	.007*
	Positive ranks	10[b]	6.00	60.00		
	Ties	42[c]				
Q2q4—Q1q4 (Attitude towards application of their previous knowledge)	Negative ranks	3[a]	2.50	7.50	−1.000[f]	.317
	Positive ranks	1[b]	2.50	2.50		
	Ties	49[c]				
Q2q5—Q1q5 (Belief about their learning)	Negative ranks	0[a]	.00	.00	−2.000[e]	.046*
	Positive ranks	4[b]	2.50	10.00		
	Ties	49[c]				
Q2q6—Q1q6 (Regarding future use of the OILE)	Negative ranks	1[a]	5.00	5.00	-.707[e]	.480
	Positive ranks	4[b]	2.50	10.00		
	Ties	48[c]				

Notes: Total N = 53. Q1—Questionnaire 1, Q2—Questionnaire 2, qi—question number
[a]Response in Q2 < Response in Q1
[b]Response in Q2 > Response in Q1
[c]Response in Q2 = Response in Q1
[d]The sum of negative ranks equals positive ranks
[e]Based on negative ranks
[f]Based on positive ranks
*p values below $\alpha = 0.05$

4.4.2 *Study II: First Stage Impact Study*

The main research question investigated in the second study was; "Does using the Mr. Wang online interactive learning environment affect the development of students' complex dynamic problem-solving skills?"

Prior to the first impact study, the equivalence of the study participants were assessed using the pre-assessment tools as discussed under the data collection section. Table 4.2 summarizes the participants' background information and the independent samples t-test results.

Table 4.2 shows that the study participants were similar regarding most of their background information. Also, the independent-samples t-test analyses, which compared the mean scores of the Experimental groups against their respective Control groups and the mean scores of the two Experimental groups against each other over the four assignments, show that there were not statistically significant differences in performances of the study participants prior to the treatment. Thus, the study participants can be considered as equivalent and any differences in performance thereafter, may be attributed to the interventions made in this research.

Following the intervention, the first stage impact study lead to two major findings. The first finding was on the impact of using the Mr. Wang OILE on the students' subsequent task performance. The students' process log information was used to assess whether scaffolding students early in their tasks, using the OILE, helped them perform better in their subsequent tasks. The results from the process log of each student demonstrate that the OILE supported the students in their learning in and about CDS. As evidenced by the paired samples t-test analysis (Table 4.3), the students' performance improved significantly across time over subsequent tasks. Effect size measurements show that the magnitude of the improvement in the students' performance, which could be attributed to the use of the Mr. OILE, is under the medium effect category (Coe, 2002).

Table 4.2 Pre-assessment results

Group		*CG N = 27	**EG1 N = 33	***EG2 N = 24	EG1 vs. CG	EG2 vs. CG	EG1 vs. EG2
Background information	Age						
	20–24	0.33	0.21	0.21			
	25–29	0.56	0.42	0.50			
	30–34	0.11	0.21	0.21			
	35 and up	0	0.15	0.08			
	Avg age	25.7	28.5	28.5			
	Gender						
	Male	0.44	0.58	0.58			
	Female	0.56	0.42	0.42			
Independent samples t-test results	Mean	7.81	7.79	7.67			
	SD	1.64	1.17	1.34			
	df				58	49	55
	t				−0.07	−0.35	0.36
	Sig (2-tailed)				0.941	0.728	0.717

Note: *CG—Control Group, **EG1—Experimental1 Group, ***EG2—Experimental2 Group

Table 4.3 Paired-samples t-test results of the Mr. Wang OILE, adopted from Tadesse and Davidsen (under review)

Group	N	Mean	SD	df	t	Sig. (2-tailed)	Effect size (Cohen's d)
Part 1: Task 1 and 2	57	7.55	0.26	56	3.28	0.001	0.43
Part 2: Task 3 and 4	57	7.46	0.26				
Part 2: Task 3 and 4	57	7.46	0.26	56	−6.00	0.000	0.76
Part 3: Task 5	57	7.62	0.22				
Part 1: Task 1 and 2	57	7.55	0.26	56	−2.39	0.019	0.30
Part 3: Task 5	57	7.62	0.22				

Note: The Mr. Wang OILE has 5 tasks divided in three parts. Part 1 consists Tasks 1 and 2, Part 2 consists Tasks 3 and 4, and Part 3 consists Task 5

Table 4.4 Independent samples t-test result of a posttest, adopted from Tadesse and Davidsen (under review)

Group	N	Mean	SD	*F (Sig.)	df	t	Sig. (2-tailed)	Effect size (Cohen's d)
Experimental1	33	8.95	1.57	12.09	40	3.03	0.004	0.8
Control	27	7.17	2.72	(0.001)				
Experimental2	24	8.60	2.12		49	2.08	0.042	0.59
Control	27	7.17	2.72					
Experimental1	33	8.95	1.57		55	0.72	0.476	0.187
Experimental2	24	8.60	2.12					

Note: *F—Levene's test for equality of variances are reported with F values and p-values under brackets on events were unequal variance were assumed

The second major finding was on the impact of using the Mr. Wang OILE on the students' problem-solving skills—which was measured using the posttest. The empirical findings from the independent samples t-test analysis (Table 4.4) show that, when scaffolded, the Experimental group students made a statistically significant improvement in their problem-solving performance compared to the Control group students, who were not scaffolded by the OILE. The effect size calculation shows that the statistically significant difference observed between the Experimental and the Control groups in the study is attributed to the use of the OILE, and the effect is under large effect category (Coe, 2002).

4.4.3 Study III: Second Stage Impact Study

The main research question investigated under this third study was; "How does scaffolding feedback, which is integrated to the Mr. Wang OILE, affect the performance of students?"

The third study had three different focus areas and lead to three major findings that were associated with each of these focus areas. The first focus area was on the impact of the scaffolding feedback on the performance gap between High- and Low-performing students. The second focus area was on the impact of the scaffolding feedback in the development of the students' transferable skills. The third and final focus area was on the association between the students' performance in the transferable skill exercise and (a) the number of feedback they received per question, (b) the amount of time they spent per feedback while working on the Mr. Wang OILE.

To assess who benefited the most from the scaffolding feedback, the Experimental group students' progress log was used. There were 57 students in the Experimental group, 33 students from the 2017 academic year and 24 students from the 2018 academic year. The Experimental group students were then divided into three subgroups based on their average performance on part one of the Mr. Wang OILE, High-, Average-, and Low-performers. Students who were within one standard deviation of the average performance, were considered as Average-performers. Those who were above one standard deviation from the average, were considered High-performers and below one standard deviation, Low-performers.

Part one of the Mr. Wang OILE had 25 question, 5 of them under Task 1 and the remaining 20 under Task 2. Task 1 focused on problem identification and Task 2 on model building. The subgrouping was done twice, the first one was based only on students' average performance in Task 1, and the second one was based on the average performance in both Task 1 and Task 2. The subgrouping was done twice to reduce the possible impact of mis-categorization of students. Students might have varying level of competences in different problem domains. However, it was assumed that the competence level in problem identification and model building can be taken as a good proxy for the students' cognitive capability of comprehending structural complexity and the skills required to infer dynamic behavior of a system from its underlying structure. Hence, in an un-supporting learning environment, it is assumed that the gap between the average performance levels of the subgroups would remain the same across the different level of competencies.

The first subgrouping resulted in 20 students, split in 10 High-performers and 10 Low-performers. The second subgrouping resulted in 23 students, 12 High-performers and 11 Low-performers. Using the progress log of these two subgroups of students, the study assessed how the gap between the High- and the Low-performers changed over time as they progressed from the first part of the OILE to the third part of the OILE, that is, from Task 1 to Task 5 of the Mr. Wang Case study—covering a total of 116 questions.

The first major finding of Study III (Table 4.5) was that the performance level of the Low-performing students increased significantly and the gap between the High- and the Low-performing students was reduced over time across subsequent tasks—as evidenced from the students' process log. In addition, the information from the students' process log show that the Low-performing students benefited the most from the scaffolding feedback compared to the High-performing students. Consequently, the Mr. Wang OILE served as a leveler (equalizer) bridging the performance gap between the High- and Low-performing students.

Table 4.5 High- and low-performer students' average performance across the five tasks of Mr. Wang case study, adopted from Tadesse (under review)

Group	Number of students	Task number	High-performers' avg	Low-performers' avg	Gap (%)
Based on Task 1 performance	20	Task 1	7.86	6.66	18.02
		Task 2	7.88	7.67	2.65
		Task 3	7.64	7.35	4.01
		Task 4	7.62	7.19	5.96
		Task 5	7.77	7.54	3.10
		Overall average gap			6.75
Based on Task 1 and Task 2 performance	23	Task 1	7.67	6.91	10.96
		Task 2	7.98	7.52	6.10
		Task 3	7.71	7.26	6.10
		Task 4	7.69	7.16	7.41
		Task 5	7.79	7.39	5.40
		Overall average gap			7.10

Table 4.6 Independent samples t-test result of the transferable skill exercise, adopted from Tadesse (under review)

Study	No. of students	Mean	SD	df	t	Sig. (2-tailed)	Effect size (Cohen's d)
Experimental2	24	13.62	5.12	49	2.07	0.044	0.58
Control	27	10.86	4.40				

Table 4.7 Pearson product-moment correlation coefficient, adopted from Tadesse (under review)

		Avg # of feedback per question	Avg time spent per feedback
Performance level	Pearson correlation	**−0.665****	0.143
	Sig. (2-tailed)	0,000	0.505
	N	24	24

**Correlation is significant at the 0.01 level (2-tailed)

The second major finding of Study III was, as evidenced from the independent samples t-test analysis (Table 4.6), the Experimental2 group students performed statistically significantly higher in the transferable skill exercises than their corresponding Control group. A detailed analysis of the students' performance level in answering individual questions, reveals that the Experimental2 group students had superior performance over their corresponding Control group on questions that require deep understanding such as describing over-time-changes in model outputs based on the underlying structures of the model. The effect size measurement confirms that the statistically significant difference observed between the Experimental2 and the Control group students was largely attributed to the scaffolding feedback provided to the students using the Mr. Wang OILE.

The third and final major finding of Study III (Table 4.7) was the identification of a correlation between the students' performance in the transferable skill exercise and (a) the number of feedback they received per question, and (b) the amount of time they spent per feedback while working on the Mr. Wang OILE. The result from the Person product-moment coefficient analysis reveals that there was a statistically significant negative correlation between the performance level in the transferable skill exercise and the number of feedback the students received per question. In other words, the students who scored the lowest on the transferable skill exercise were those who were receiving the highest number of feedback per question.

There could be two possible explanations for the negative correlation: (1) It might be trivial to think that academically weaker students could demand more feedback compared to the stronger students. However, it would be difficult to think that the weaker students could outperform the stronger students due to the support they received by way of the OILE so that the correlation could be positive. (2) During the design of the Mr. Wang OILE, it was decided that all students, including those who responded correctly, should receive Knowledge of concept feedback as part of the scaffolding feedback—feedback that informs the students why their answers were correct and why the other alternatives were wrong (see Chap. 3). Thus, even if the High-performing student did make fewer mistakes and, consequently, received less feedback, they still had a good chance to learn from the Knowledge of concept feedback and kept performing well in the transfer skill exercises. Hence, it is probable that the design of the scaffolding feedback might have confounded with the observed result.

The analysis on the correlation between the students' performance and the amount of time they spent per feedback reveals a statistically insignificant positive correlation. Hence, the study failed to reject the null hypothesis, which claimed there is no association between the amount of time the students spent per feedback and their performance in the transferable skill exercise.

Appendix 1: Mr. Wang's Bicycle Repair Shop Case Study

An adaptation of the Causes of Oscillation exercise published by the System Dynamics Group, MIT, Cambridge, Massachusetts, USA.

Introduction

In Shanghai, Mr. Wang owns a bicycle repair shop called *Mr. Wang*, that was established by his great grandfather in 1912. The repair shop is the most reputable one in Shanghai and has always prided itself of a one-day service time. The core workforce in Mr. Wang consists of members of his family.

Mr. Wang has recognized a dynamic problem and was recommended to turn to a consulting company, *iBelieve*, for assistance. After *iBelieve* gave up the task, Mr.

Wang has asked the Masters Class in System Dynamics at University of Bergen for assistance. He states his problem as follows:

Mr. Wang has a relatively constant client base. But Shanghai is a very dynamic environment with a lot of events—be they of commercial, athletic, or other sorts. Ever so often, therefore, there is an inflow of people to Shanghai that prefer to use bicycles to get around. Bike rental is very popular and the repair business is booming. The problem is that, due to the events, each increase in repairs takes place discretely (at a point in time)—not continuously. And these single events have effects on *Mr. Wang* of an unexpectedly long duration.

Mr. Wang demonstrates his problem by showing us some data from three events over the last 2 years (units on time axis is days), occurring at day 10, day 310 and day 410, and having repercussions into year 2 (Fig. 4.1):

Having filtered out noise from a random order rate, the system will behave like the graph shown in Fig. 4.2.

Each event lead to the introduction of an additional 3–5% bicycles in Shanghai, predominantly supplying the rental market. And because of Mr. Wang's prominence, the company experienced a permanent 10% increase in orders (demand for repairs) after each such event. One of the reasons for this increase is that people who rent bikes are not as careful in handling and maintaining their bikes as the regular citizen of Shanghai. A 10% increase in orders is not very much and one would

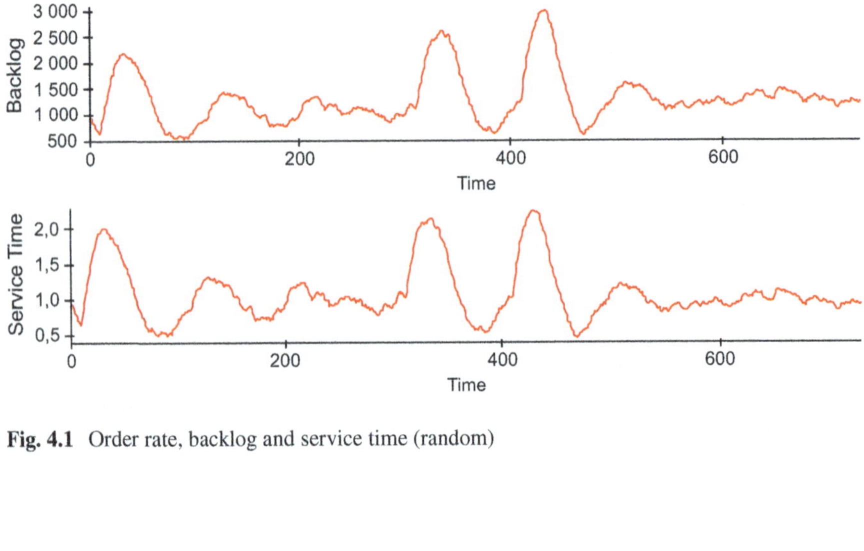

Fig. 4.1 Order rate, backlog and service time (random)

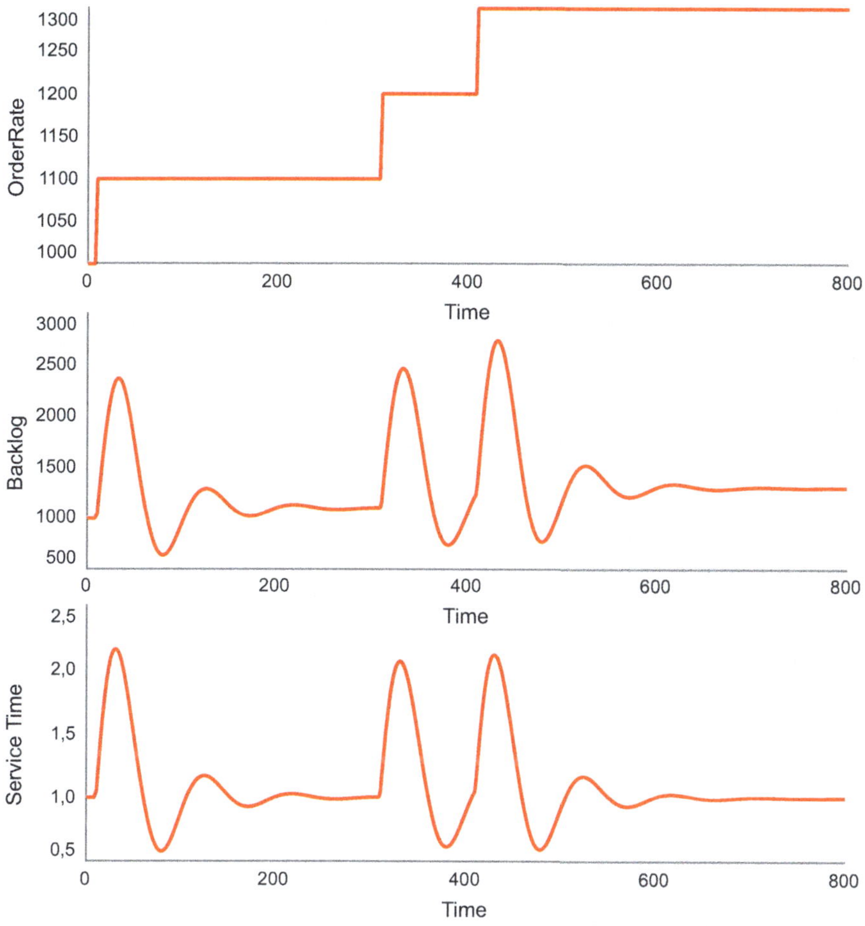

Fig. 4.2 Order rate, backlog and service time (deterministic)

perhaps expect that such an increase could be handled easily by a company like Mr. Wang. Alas, not so! In fact, the Backlog rises by around 150% right after each event only to return below target and oscillate for a considerable time after each event! And so does the service time!

In the short run, therefore, it seems that the company is not able to tackle the increase in demand very well. In the very long run, there are signs that indicate the company works to the satisfaction of Mr. Wang—stabilizing the Backlog and the Service Time effectively.

Appendix 2: Mrs. Lee's Bicycle Factory Case Study

An adaptation of the Causes of Oscillation exercise published by the System Dynamics Group, MIT, Cambridge, Massachusetts, USA.

Introduction

In Beijing, Mrs. Lee owns a bicycle shop Mrs. Lee, that was established by her great grandfather in 1922. The bicycle factory is the most reputable one in Beijing and has always prided itself of being able to deliver on time—to bicycle shops and rental firms across the city.

Mrs. Lee has recognized a dynamic problem and was recommended to turn to a consulting company, *iBelieve,* for assistance. After *iBelieve* provided some advice that turned up not to yield the satisfactory results, Mrs. Lee has asked the Masters Class in System Dynamics at University of Bergen for assistance. She states her problem as follows:

Mrs. Lee has a relatively constant client base. But Beijing is a very dynamic environment hosting a lot of events—be they of commercial, athletic, or other sorts. Ever so often, therefore, there is an inflow of people to Beijing that prefer to use bicycles to get around. Bike sales and rentals are very popular, and the factory is booming. The problem is that, due to the events, each increase in demand for bikes takes place discretely (at a point in time)—not continuously. And these single events have some adverse effects for Mrs. Lee both in the short and in the long run.

Mrs. Lee demonstrates her problem by showing us some data from three events over the last 2 years (units on time axis is days), occurring at day 10, day 310 and day 410, and having repercussions into year 2 (see Fig. 4.3).

The lost sales over this period is in the order of 1500 (1490). Filtering out noise, the system will behave like this (Fig. 4.4):

Each event leads to the demand for an additional 10% bicycles in Beijing, predominantly originating from the rental market. A 10% increase in orders is not very much and one would perhaps expect that such an increase could be handled easily by a company like Mrs. Lee. Alas, not so! In fact, the Inventory approaches 0—often reaching that level, with the result that the customers are not being served as expected and that sales are lost. In the long run, this may cause a loss in customers and the associated demand.

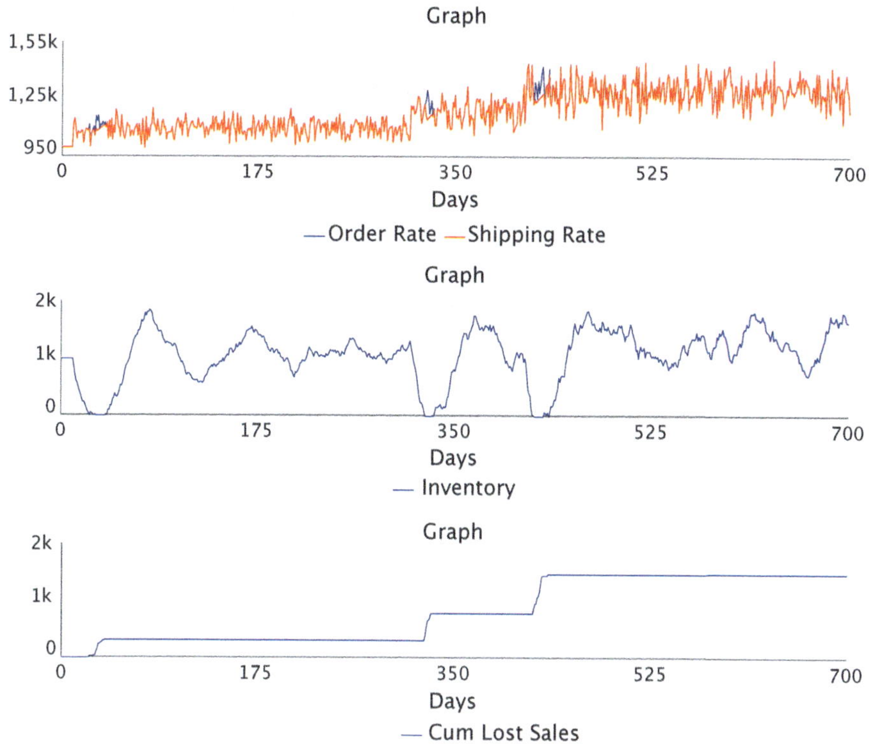

Fig. 4.3 Order (real) and shipping rates, inventory and cumulative lost sales

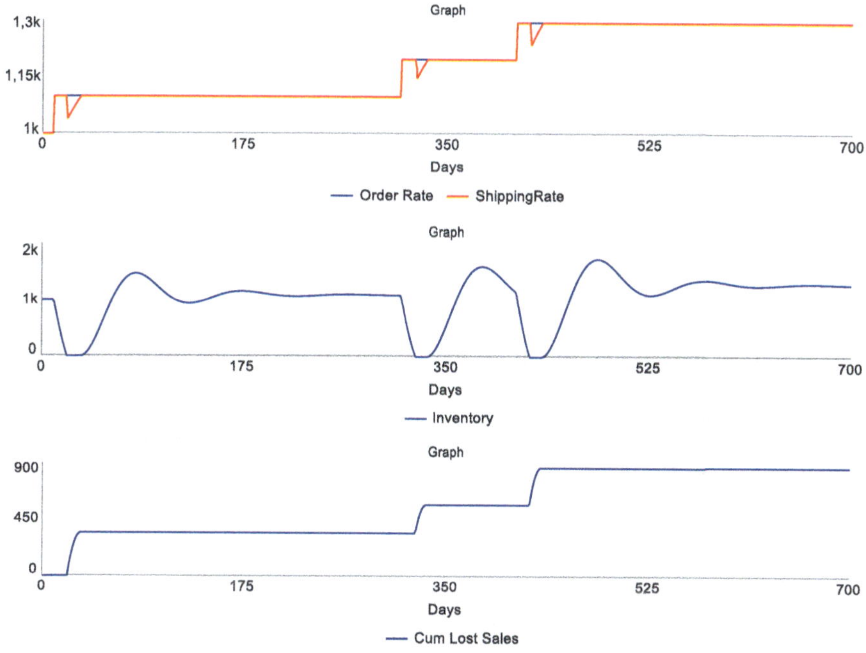

Fig. 4.4 Order (average) and shipping rates, inventory and cumulative lost sales

Appendix 3: Students' Response to Questionnaires

Table 4.8 Students' response to Questionnaire 1 and 2, adopted from Tadesse and Davidsen (2019). Due to rounding, total sums could be different from 100%

	Questions	Strongly disagree	Disagree	Neither agree/ disagree	Agree	Strongly agree
1	Experience with the user interface of the OILE: It has clear interface It is easy to navigate through the OILE The OILE does not have unnecessary long texts that hinder my learning	1	11	13	57	18
2	Attitude towards the content of the learning and its organization: It is appropriate to students of my level I have learned from the tasks It helped me learn step by step	1	5	17	54	23
3	Attitude towards the feedback offered: I have read all the feedback I have learned from the feedback	3	18	15	53	11
4	Attitude towards application of the knowledge and skills they acquired in a previous course: The OILE gave me the opportunity to practice the skills I gained during a previous course	1	6	13	68	12
5	Belief about their learning: I have understood the objective of the case study I have understood the main problem in the case study I have gotten deeper insight about the main concepts of the case I am ready to embark on the next challenge of the course	1	4	18	55	22
6	Regarding future use of the OILE: I recommend other system dynamics students of my level to make use the OILE	1	0	15	46	38

Notes: N = 57, response rate 53/57 = 93%

References

Anderson, T., & Shattuck, J. (2012). Design-based research: A decade of progress in education research? *Educational Researcher, 41*(1), 16–25.

Babbie, E. (2015). *The practice of social research* (14th ed.). Wadsworth/Thomson.

Berkeley Center for Teaching & Learning. (n.d.). *Course evaluations question bank.* Retrieved August 10, 2017, from https://teaching.berkeley.edu/course-evaluations-question-bank#anchor3

Coe, R. (2002). *It's the Effect Size, Stupid: What effect size is and why it is important.* Paper presented at the Annual Conference of the British Educational Research Association. University of Exeter, England.

Creswell, J. W., & Creswell, J. D. (2018). *Research design; Qualitative, quantitative, and mixed methods approaches* (5th ed.). Sage.

Davidsen, P. I. (1996). Educational features of the system dynamics approach to modelling and simulation. *Journal of Structural Learning, 12*(4), 269–290.

Fowler, F. J. (2014). *Survey research methods* (5th ed.). Sage.

Ifenthaler, D., & Eseryel, D. (2013). Facilitating complex learning by mobile augmented reality learning environments. In R. Huang, S. Kinshuk, & J.M. (Eds.), *Reshaping Learning: Frontiers of learning technology in a global context* (pp. 415–438). Springer.

Johnson, R. B., Onwuegbuzie, A. J., & Turner, L. A. (2007). Toward a definition of mixed methods research. *Journal of Mixed Methods Research, 1*(2), 112–133.

Jonassen, D. H. (2000). Toward a design theory of problem solving. *Educational Technology Research and Development, 48*(4), 63–85.

Maor, D., & Fraser, B. J. (2005). An online questionnaire for evaluating students' and teachers' perceptions of constructivist multimedia learning environments. *Research in Science Education, 35*(2–3), 221–244.

McKenney, S., & Reeves, T. C. (2013). Systematic review of design-based research progress: Is a little knowledge a dangerous thing? *Educational Researcher, 42*(2), 97–100.

Sheskin, D. J. (2003). *Parametric and nonparametric statistical procedures* (3rd ed.). Chapman & Hall/CRC.

Spector, J. M., & Anderson, T. M. (2000). *Integrated and holistic perspectives on learning, instruction and technology.* Kluwer Academic Publishers.

Tadesse, A. T. (under review). Scaffolding feedback in complex dynamic system context: Effect of online interactive learning environments. Manuscript submitted for publication in *Technology, Knowledge and Learning.*

Tadesse, A. T., & Davidsen, P. I. (2019). Framework to support personalized learning in complex systems. *Journal of Applied Research in Higher Education, 12*(1), 57–85.

Tadesse, A. T., & Davidsen, P. I. (under review). Problem solving skills in complex dynamic system context: Effect of online interactive learning environments. Manuscript submitted for publication in *System Dynamics Review.*

Taylor-Powell, E., & Renner, M. (2009). *Collecting evaluation data: End-of-session questionnaires*, University of Wisconsin—Extension, Cooperative Extension, Program Development and Evaluation, Madison, Wisconsin.

van Merriënboer, J. J., & Kirschner, P. A. (2017). *Ten steps to complex learning: A systematic approach to four-component instructional design.* Routledge.

Chapter 5
Lessons for Practice and Conclusion

Research shows that we experience a multifaceted challenge when we try to understand and communicate our understanding about complex, dynamic systems. The challenges emanate from three different sources, structural complexity of dynamic systems, lack of skills to understand and communicate our understanding about such system, and the level of effectiveness of methods, techniques, and tools that are available to support and measure our understanding. The authors of this monograph aimed at enhancing students learning in and about CDS by designing a personalized and adaptive OILE based on existing theories and methods that consist of a well-composed set of instructional techniques. In the research, the techniques were manifested in the form of an educational tool, the Mr. Wang OILE, to support the students in their study of CDS. This chapter discusses the overall implication of the research and its major contribution focusing on practical, theoretical, and methodological implications. Also, the chapter presents a summary of key instructional design principles, the limitations of the research, recommendations for future research, and concluding remarks.

5.1 Practical Implication

Spector (2017), in his summary of the 1990s debate between Richard Clark and Robert Kozma regarding the influence of media in learning, argues that:

> Media and technology can provide affordances and possibilities not previously available, but effective use of media and technology was still dependent on good instructional design as well as training and support for those using the technologies. What makes an instructional design good? Remember the goal—help people learn. An effective instructional design is one that can be demonstrated to have a positive impact on learning. (p. 1419)

In line with Spector's view, the two main practical contributions of this monograph are increased knowledge about how to effectively design a personalized and

A. T. Tadesse et al., *Adapting Interactive Learning Environments to Student Competences*, SpringerBriefs in Educational Communications and Technology,
https://doi.org/10.1007/978-3-030-88289-1_5

adaptive OILE and increased knowledge about the positive impact of the use of the OILE in the students' learning in and about CDS.

First and foremost, following a design-based research (Barab & Squire, 2004; McKenney & Reeves, 2012, 2013; Huang et al., 2019) the monograph demonstrates how an effective and practical personalized and adaptive online interactive learning environment can be designed and tested in four steps. First, it shows how a practical educational problem, difficulties of learning in and about CDS, can be identified by exploring relevant literature and conducting focus group discussions with experts. Second, the research shows how such a practical educational problem can be addressed by designing effective instructional methods, which consists of well-composed set of instructional techniques, based on existing instructional theories and/or models. The instructional techniques were manifested in the form of an educational tool, the Mr. Wang OILE. Third, the research showcases how the designed instructional tool can be implemented and tested iteratively. Finally, the research shows how reflections and design principles can be formulated.

The other practical contribution of the research associated with the design of the Mr. Wang OILE is the effective integration of dynamic content delivery, evaluation of performance, provision of support, and collection of process log data under a single online interactive learning environment. The Mr. Wang OILE has been designed to deliver the content of the learning material dynamically adapting to individual student's performance level. Although the OILE has been designed based on "a static model of the content to be learned and a static model of common misconceptions and misunderstanding" (Spector, 2018, p. 39), it delivers the content dynamically based on the stage and performance level of the individual student. To deliver the content dynamically, the OILE does a performance evaluation so that it can choose the proper support at the right time. While all these processes take place, the OILE records the students' learning analytics in the form of a process log. The integration and provision of all these processes and activities under a single OILE can be considered as one of the practical contributions of this monograph. Since this practical contribution also demonstrates how to integrate the above-mentioned processes and activities systematically in the OILE, it can be considered as a methodological contribution as well.

The second practical contribution of the research is increased knowledge about the impact of personalized and adaptive OILE on the students' learning in and about CDS. Spector (2018) defines learning as "stable and persistent change in what a person knows, believes, and/or can do" (p. 1421). In line with Spector's definition, the monograph provides evidence on how the use of the OILE positively affected the students' belief about their learning, and in what they know and can do across time over different tasks of the Mr. Wang OILE.

Arranged in a blended learning setup (Means et al., 2010), the Mr. Wang OILE was tested with the System Dynamics master program students at the University of Bergen during the fall semesters of 2017 and 2018. The study shows that the use of the Mr. Wang OILE positively impacted five different aspects of the students' learning. The OILE had positive impacts (1) in the students' affective domain of learning, (2) in their problem-solving skills, (3) in their transferable skills, (4) in their deep understanding, and (5) in bridging performance gap between the High- and the Low-performing students.

Study I of the monograph (see Chap. 4) reports about the students' attitude *towards* and their experience *with* the Mr. Wang OILE. The analysis of the two questionnaires, which were administered to the students during and after they completed working on the Mr. Wang OILE, shows that the students were through an effective learning experience and they had strong confidence in their own learning.

Study II of the monograph demonstrates how the students problem-solving skills have improved. The findings in Study II show that the students' performance on the Mr. Wang OILE improved statistically significantly over time across different tasks as evidenced by the students' process log and the paired samples t-test analysis. In addition, Study II reveals that when scaffolded by the Mr. Wang OILE, the students made a statistically significant improvement in their problem-solving compared to those who were not scaffolded.

Study III of the monograph shows that the use of the Mr. Wang OILE helped to reduce the performance gap initially observed between the High- and the Low-performing students, mainly due to a significant increase in the performance level of the Low-performing students. In other words, the Mr. Wang OILE served as an equalizer. Furthermore, Study III demonstrates that students who used the Mr. Wang OILE performed significantly higher than their corresponding Control group on the transferable skill exercise and their performance was significantly higher on questions that require deep understanding.

Overall, the research shows that using the Mr. Wang OILE to support students' learning in and about CDS brought positive change in both the students' affective and cognitive domains of learning. Moreover, the findings of the three studies accounted the major findings of more than 50 years of research on learning. Those three findings are first that prior performance tends to predict future performance, which implies potentially different scaffolding and feedback for Low- and High-performing students. Second, timely and informative feedback tends to enhance performance. Third, time on task tends to predict performance.

The third practical contribution of the monograph is an increased knowledge regarding the affordance of personalized and adaptive OILE to teachers and/or mentors and to students as well.

The use of the OILE helps teachers and mentors to identify students that needed extra support. Some students are very shy to come forward and ask for help either from their teachers or from their colleagues due to cultural and experiential perspectives. At the same time, it is difficult for a teacher to identify students who are experiencing difficulties with their learning material unless either the students go to the teacher on their own or their exam results are published. However, the presence of reports about individual student's learning path following their use of the Mr. Wang OILE helps to identify those students that need extra remedies. Moreover, such reports also help to identify topic areas that need extra discussion during face-to-face session, particularly those topic areas where large number of students repeatedly failed to answer correctly, were identified for further action.

For the students, the Mr. Wang OILE gave them the opportunity to do hands-on computer modeling activities, while they were receiving immediate feedback about their model structures and model outputs. In the Mr. Wang OILE, there are questions that ask students to build model structures in their own personal computers and report the model structures and model outputs comparing their own works with what have been asked in the OILE. Such questions and the associated feedback communicated to the students following their response, helped them carry out hands-on modeling activities. Moreover, using the Mr. Wang OILE allowed the students to recognize their own performance level. It also gave them a chance to move back and forth between topic areas they learned already in a previous System Dynamics course and those they were learning in the new course, the course of which the Mr. Wang case study is a part. As described before, the Mr. Wang OILE has been designed to adapt to individual student's performance level. Hence, when a student fails to correctly answer a question, the OILE provides an opportunity for the student to branch to a question that is conceptually easier than the first question. In doing so, a student can branch to a level where the foundational concepts of the previous System Dynamics course have been presented. Practically, the OILE has been designed to help the students connect the concepts of the two courses during their study.

5.2 Theoretical Implication

The first theoretical contribution of this monograph is the use of the holistic perspective both in the selection of instructional design models (Spector & Anderson, 2000; van Merriënboer & Kirschner, 2017) and design of learning materials (Merrill, 2002, 2013; Francom & Gardner, 2014; Francom, 2017).

The main objective of the research is enhancing students' learning in and about CDS, i.e., fostering a systems thinking (holistic) perspective. The systems thinking approach is *the whole is always more than the sum of its parts*, indicating that the structure of parts synergizes to produce the resulting dynamics of a system. Hence, the instructional design models considered during the design of the Mr. Wang OILE were those that can foster this holistic perspective.

van Merriënboer and Kirschner (2017) argue that using the holistic instructional design approach during the design of an instruction helps to address "three persistent problems" in the field of education: "compartmentalization, fragmentation, and transfer paradox" (p. 5). van Merriënboer and Kirschner claim that instructional models that apply the holistic perspective, do not compartmentalize domains of learning focusing only either on cognitive, affective, or psychomotor domains of learning. Rather, they consider the different learning domains all together. Similarly, such instructional design models do not support the practice of breaking general learning objectives into small and incomplete or isolated parts. Also, they give due consideration to the transfer of knowledge and skills by reducing focus on highly specific objectives.

This monograph showcases how the six instructional design models that promote the holistic instructional design, have influenced the development of the Mr. Wang OILE; the 4C/ID, the First Principles of Instruction, the CLE, the TCI, the Cognitive Apprenticeship, and the Elaboration Theory. In addition, the research provides evidence regarding how the issues of compartmentalization, fragmentation, and transfer paradox were addressed with the use of the holistic instructional design approach. Th monograph demonstrates how the affective and cognitive domains of learning were considered in this research while focus was also given to transferable skills, and to broader learning objectives that range from problem conceptualization to model analysis and policy design.

The holistic instructional design approach in general and the six instructional design models that influenced the development of the Mr. Wang OILE offer, in particular, unifying perspectives regarding the choice and design of learning materials. They suggest that the learning tasks should;

- be at the center of the instructional design;
- be based on authentic problems;
- comprise the entire knowledge and skills that learners would be able to acquire when they complete the entire learning tasks;
- be designed in a way that learners can address the authentic problem in its entirety, from "start to finish, rather than discrete pieces" of the problem;
- be designed in a way that learners can progress from simple to complex steps in their analysis of the entire task.

These guiding principles were considered in this research during the choice and design of the Mr. Wang and the Mr. Lee case studies, signifying the theoretical contribution of the research in terms of effectively utilizing the holistic perspective in the choice and design of learning tasks.

The second theoretical contribution of this research is the use of the instructional scaffolding method (Wood et al., 1976; Belland, 2017) during the design of instructional methods and instructional techniques to support learning in and about CDS using the Mr. Wang OILE. Wood et al. (1976) define scaffolding as a "process that enables a child or novice to solve a problem, carry out a task or achieve a goal which would be beyond his unassisted efforts" (p. 90). The support provided is "meant to extend students' current abilities" so that they can carry out the "bulk of the work required to solve the problem" (Belland, 2017, p. 17).

The instructional scaffolding method has been widely applied in the STEM fields and is found very effective especially in enhancing the students' cognitive domains of learning (Belland, 2017). As discussed in Chaps. 2 and 3, the instructional scaffolding method comprises three elements: dynamic assessment, provision of just the right amount of support, and intersubjectivity (Belland, 2017). The monograph showcases how the instructional scaffolding method has been used as the core instructional method in the Mr. Wang OILE to support the students learning in and about CDS and how the three elements of the scaffolding method have been used in the design of the instructional techniques of the Mr. Wang OILE.

In one of the widely cited educational article, Kirschner et al. (2006) signify the importance of providing guided instruction while criticizing the absence or the low level of guidance provided to students when instructional methods such as constructivist, discovery, problem-based, experiential, and inquiry-based methods are applied. Kirschner and his colleagues assert that unless the students have "sufficiently high prior knowledge" that would provide them "internal guidance", they should receive guidance that would enable them to successfully complete the learning material with the desired level of understanding. This monograph provides evidence how the use of the instructional scaffolding method and its associated instructional techniques in the OILE supported the development of the students' problem-solving skills and the transferable skills in their study of CDS.

The third theoretical contribution of the research is the application of both the holistic and scaffolding perspectives in the design of the scaffolding feedback. Feedback practitioners and learning scientists connect the notion of educational feedback with the concept applied in cybernetics, which deals with control of systems, though the actual practice often found short of this notion (Shute, 2008; Boud & Molloy, 2013). Wiener (1954) described the notion of feedback in cybernetics as:

> Feedback is a method of controlling a system by reinserting into it the results of its past performance. If these results are merely used as numerical data for the criticism of the system and its regulation, we have the simple feedback of the control engineers. If, however, the information which proceeds backward from the performance is able to change the general method and pattern of performance, we have a process which may well be called learning. (p. 61)

This notion of feedback implies that, when an educational feedback is communicated to students, either through human agent or through programed systems such as the OILE, that it should not merely aim at telling the students about their performance level, as widely seen in the actual feedback practice (Boud & Molloy, 2013). Rather, it should aim at helping them realize the gap between their current performance and the desired performance and support them to change from their current state of understanding so that learning can occur. This is the very basic notion of scaffolding described above. But to achieve this, the feedback process needs to consider the whole system. It needs to consider the state of the student (the students current performance level), the desired (standard) level of performance, the gap between the current and the desired level of performance, the right information that need to be communicated to alter the gap, and, finally, a mechanism for checking whether the desired change has been achieved or not. For the student to learn from the information communicated and for that information to be considered as feedback, all the components of the feedback process need to work together in interaction, which is the concept of the holistic perspective.

This monograph provides evidence on how one can apply the original notion of feedback adapted from the cybernetics by showcasing the design of the scaffolding feedback and how the use of such feedback can practically affect the students learning.

In the scaffolding feedback presented in this research, before students work on a specific item or task, the OILE presents a context or feed-up information that demonstrates the learning goal and the desired performance level for that specific item. Then their performance in the item is assessed in comparison with the desired performance level and then either a feed-back or feed-forward is provided to the students. The feed-back communicates to the students their performance gap and provides information that would help them fill the gap. Whereas, the feed-forward acknowledges the students correct performance and provides the reason why their responses are correct and why the other alternatives are not correct.

5.3 Methodological Implication

The use of the design-based research (Barab & Squire, 2004; Reeves, 2006; Herrington et al., 2007; Anderson & Shattuck, 2012; McKenney & Reeves, 2012, 2013; Huang et al., 2019) together with the mixed methods research (Johnson et al., 2007) can be considered as the first methodological contribution of this monograph. Motivated by the recommendation of Herrington et al. (2007) and other researchers in the field, the authors of this monograph applied the design-based research as the overarching research design. The research benefited from the design steps and procedures suggested in the literature for both the DBR and the mixed methods research. The steps followed in this research and the implications drawn from the research, demonstrate the richness of the chosen design and research methods in carrying out educational technology research. The authors believe this monograph can be considered as a methodological showcase to demonstrate the application of both the DBR and the method methods research.

The second methodological contribution of the monograph is the use of process log as a source data for the research. This innovative data collection strategy helped to collect rich data about individual students' activities while they were studying CDS. The richness of the data helped inform the development of the students learning paths, study the amount of support provided to each and every student during their study, and carry out analysis on the performance of both individual and the whole group of students at each stage of the learning material.

The third and final methodological contribution of the monograph is transparency. The main guiding principle that has been followed throughout the research process is promoting transparency of research so that other educational technology researchers that have the mission of enhancing students learning in and about CDS can replicate or utilize the research design and methods in a similar or related context. Throughout the monograph the research design, the methods for data collection, and the frameworks for data analysis were thoroughly described. Moreover, the instruments used for data collection and the results were properly documented.

5.4 Summary of Key Instructional Design Principles

Based on findings and lessons learned from the research process of this monograph, the following seven key instructional design principles are proposed for consideration when designing online interactive learning environments that support learning in and about complex, dynamic systems:

1. Consider using the five-step design framework of this research that starts with the identification of instructional design models and ends with the implementation of the learning environment.
2. Consider using design-based research as the overarching design of the research.
3. Consider sequencing learning materials from simple to complex.
4. Consider providing scaffolding feedback that fades away as learners gain expertise while students are studying CDS.
5. Consider measuring students' performance and tracking their process log to generate learning analytics that would help improve the learning environment in the future, identify students that need extra support, and adjust and improve the face-to-face session.
6. Consider measuring both the affective and cognitive domains of learning.
7. Consider measuring students' transferable skills after their experience with the learning environment.

5.5 Limitations and Recommendation for Future Studies

Similar to any research activity, the research carried out under this monograph has its own limitations. This section presents the limitations of the research and issues that need further research.

The monograph addresses specifically the provision of support to individual students during their study in and about CDS and hence, does not address issues associated with collaborative learning. Most of the existing platforms that support collaborative learning in and about CDS focus on the dynamics of the groups' interaction without offering detailed account for the individual students need. Future studies need to address how to foster collaborative learning in and about CDS while accounting for individual student's need as well.

Another limitation of the research is associated with the assessment instruments used in the Mr. Wang OILE. The Mr. Wang OILE relied heavily on the use of multiple-choice questions and on open-ended questions, which ask students to draw (estimate) the over-time development of variables that have significant impact on the Mr. Wang's problem formulation. However, future work should consider diverse assessment instruments such as essay type questions and questions that address the students' reflective and comprehension skills.

The third limitation of the research is its inability to generate reports automatically that are easy to read and to interpret. Except reports about the students' estimate for the over-time developments of variables, the students' process log was collected first in the form of csv files and then manually converted into spreadsheets before the data was coded into learning paths with the help of the GraphViz software (http://graphs.grevian.org/graph) and the Stella Architect software. The students' estimate for the variables over time development can be automatically generated in both time series graphs and CVS files. Given the advancements in artificial intelligence and computer technology, future studies should consider the automatic generation of important reports such as the students learning paths that are easy to read and interpret.

The fourth limitation that it is worth mentioning, is associated with the educational media used in the Mr. Wang OILE. The research is limited to the use of simulations, texts, and graphs. Inclusion of additional educational media such as audios and videos might potentially increase the learners' active engagement with the OILE.

One other limitation of the research, particularly related to Study II and Study III, is the non-random selection and assignment of samples to the two quasi-experimental studies carried out in the studies. Under such experimental design, it could perhaps be difficult to generalize the results achieved from the two studies to the larger population.

It is also important to mention here that a design limitation in the transferable skill exercise in Study III might have confounded the observed result and the conclusions made. As described previously, the Experimental2 students did the Mrs. Lee case study after they completed the Mr. Wang OILE in paper and pencil format, whereas, the Control group did only the Mr. Wang case study in paper and pencil format. Then the two groups performances were compared on identical questions that were available in both the Mr. Wang & Mrs. Lee case studies. However, the fact that the Experimental2 students did the first part of the Mr. Wang case study using the OILE might have provided them extra advantage when they work on the Mrs. Lee case study and the observed result might have been confounded due to such test design limitations. Future studies may consider allowing the Control group students to do the Mrs. Lee case study as well, and then study the difference in performance between the treatment groups.

Spector (2018), in his commentary regarding smart learning environments potential and pitfalls, notes that "there are indeed many possibilities for smart technologies to improve learning and instruction; however, … these possibilities have yet to be realized on a large scale and sustained beyond the efforts of demonstration projects" (p. 34). This research shares Spector's concern and calls for further studies and large-scale developments of personalized and adaptive OILE to support learning in and about CDS beyond demonstration studies.

In the System Dynamics Group, University of Bergen, an ERASMUS+ project, funded by the EU, has been initiated aimed at developing, in collaboration with their European university partners, System Dynamics MOOCs (Massive Open Online Courses). This development is based on lessons learned from this research, from a MOOC previously developed[1] by the SD Group, and from the latest developments in the area.

[1] An undergraduate level web-based course has been developed by the System Dynamics Group, University of Bergen to teach students about natural resource management (Alessi et al., 2012).

5.6 Conclusion

Research shows the world is facing a wide range of increasingly complex dynamic problems in both the public and private sectors; climate change, unemployment, health problems, famine, migration, etc. create challenges for private and public organizations (OECD, 2017). These problems are often dynamic (i.e. develop over time) and they commonly originate from the internal structure of the systems with which the problems addressed are associated (Diehl & Sterman, 1995; Davidsen, 1996). The structure of a system is made up of the cause and effect relationships that exist between the attributes (variables) that define a system. And the complexity of a system is defined by the diversity of that systems structure.

The main objective of this monograph was to enhance students learning in and about complex, dynamic systems by developing effective instructional methods, techniques, and tools; so that students can develop deep intuitions about complex, dynamic systems, and an ability to reveal quick fixes that ignore real world complexity (OECD, 2017; Sterman, 2011). For that purpose, a personalized and adaptive online interactive learning environment was proposed to be developed on the bases of a five-step holistic instructional design framework. The five steps of the design framework are: (1) Identification of instructional design models, (2) Identification of authentic learning material, (3) Identification of instructional methods, (4) Identification of instructional techniques, and (5) Design of the interface and implementation of the tool.

Six instructional design models influenced the development of the OILE: 4C/ID, First Principles of Instruction, CLE, TCI, Cognitive Apprenticeship, and Elaboration Theory. The OILE has the following three characteristics:

1. It presents an authentic, complex dynamic problem that the learner should address in its entirety. It then proceeds to allow learners to progress through a sequence of gradually more complex learning tasks.
2. It allows for the learner to interact with the OILE while solving the problem at hand. Upon the completion of each learning task and based on their individual performance, the OILE provides the learners with information intended to facilitate the learning process. The support fades away as learners gain expertise.
3. It tracks and collects information on students' progress and generates learning analytics that are being used to assess students' learning and to tailor the information feedback to the students.

Following a design-based research and mixed methods research, the OILE was practically implemented with an authentic case study, the Mr. Wang Bicycle Repair Shop case study, which was designed to teach master program students at the University of Bergen about the causes of oscillation (major disturbances) in a complex and dynamic system.

"The course is open to students worldwide and offers 10 credit points" (https://www.uib.no/en/rg/dynamics/50295/natural-resources-management).

A survey study and two experimental impact studies were conducted to assess the effectiveness of the Mr. Wang Bicycle Repair Shop OILE, named after the case study, in enhancing the students' learning. The studies aimed at assessing impacts both on the students' affective and cognitive domains of learning. The studies were conducted with three cohorts of System Dynamics master program students at the University of Bergen over a three-year period, from 2016 to 2018. Eighty-four students were involved in the study.

In the survey study, two questionnaires with 38 questions were administered to the students. The experimental studies were carried out using the students' process log data from the Mr. Wang OILE and their performance on a posttest and a transferable skill exercise administered after the students' used the OILE.

Analyses of the two questionnaires show the students firmly believe they have been through an effective learning experience while working within the Mr. Wang OILE. Findings from the experimental studies show that, when scaffolded using the OILE, students made a statistically significant improvement in their problem solving, which was measured using the posttest, compared to those who were not scaffolded. Results from the students' process log demonstrate that the students' performance improved significantly across time over subsequent tasks. In addition, findings from the process log show that the performance level of Low-performing students increased significantly and the gap between High- and Low-performing students reduced across time over subsequent tasks. Consequently, the Low-performing students benefited most from the scaffolding feedback compared to the High-performing students. Results from the transferable skill exercise show that students who used the Mr. Wang OILE performed significantly higher than those who did not. Effect size measurements, carried out in this study, confirm that the observed statistical differences between the treatment groups were largely attributed to the use of the Mr. Wang OILE. In light of supportive evidence, the authors of this monograph concludes that the use of OILE to support learning in and about CDS is effective and promising. Consequently, we call for further studies and large-scale developments of personalized and adaptive OILEs to support learning in and about CDS beyond demonstration studies such as this monograph.

References

Alessi, S., Kopainsky, B., & Moxnes, E. (2012). Teaching resource management with web-based models and multi-player games. In *Proceedings of the 30th international conference of the system dynamics society*. St. Gallen, Switzerland.

Anderson, T., & Shattuck, J. (2012). Design-based research: A decade of progress in education research? *Educational Researcher, 41*(1), 16–25.

Barab, S. A., & Squire, K. (2004). Design-based research: Putting a stake in the ground. *The Journal of the Learning Sciences, 13*(1), 1–14.

Belland, B. (2017). *Instructional scaffolding in STEM Education: Strategies and efficacy evidence*. Springer Open.

Boud, D., & Molloy, E. (2013). Rethinking models of feedback for learning: The challenge of design. *Assessment and Evaluation in Higher Education, 38*(6), 698–712.
Davidsen, P. I. (1996). Educational features of the system dynamics approach to modelling and simulation. *Journal of Structural Learning, 12*(4), 269–290.
Diehl, E., & Sterman, J. D. (1995). Effects of feedback complexity on dynamic decision making. *Organizational Behavior and Human Decision Processes, 62*(2), 198–215.
Francom, G. M. (2017). Principles for task-centered instruction. In C. M. Reigeluth, B. J. Beatty, & R. D. Myers (Eds.), *Instructional design theories and models: The learner-centered paradigm of education* (Vol. 4, pp. 65–91). Taylor & Francis.
Francom, G. M., & Gardner, J. (2014). What is task-centered learning? *TechTrends, 58*(5), 27–35.
Herrington, J., McKenney, S., Reeves, T. C., & Oliver, R. (2007). Design-based research and doctoral students: Guidelines for preparing a dissertation proposal. In *Proceedings of world conference on educational multimedia, hypermedia and telecommunications* (pp. 4089–4097). AACE.
Huang, R., Spector, J. M., & Yang, J. (2019). Design-based research. In *Educational technology. Lecture notes in educational technology*. Springer.
Johnson, R. B., Onwuegbuzie, A. J., & Turner, L. A. (2007). Toward a definition of mixed methods research. *Journal of Mixed Methods Research, 1*(2), 112–133.
Kirschner, P. A., Sweller, J., & Clark, R. E. (2006). Why minimal guidance during instruction does not work: An analysis of the failure of constructivist, discovery, problem-based, experiential, and inquiry-based teaching. *Educational Psychologist, 41*(2), 75–86.
McKenney, S., & Reeves, T. C. (2012). *Conducting educational design research*. Routledge.
McKenney, S., & Reeves, T. C. (2013). Systematic review of design-based research progress: Is a little knowledge a dangerous thing? *Educational Researcher, 42*(2), 97–100.
Means, B., Toyama, Y., Murphy, R., Bakia, M., & Jones, K. (2010). *Evaluation of evidence-based practices in online learning: A meta-analysis and review of online learning studies*. US Department of Education.
Merrill, M. D. (2002). A pebble-in-the-pond model for instructional design. *Performance Improvement, 41*(7), 41–46.
Merrill, M. D. (2013). *First principles of instruction: Identifying and designing effective, efficient and engaging instruction*. Pfeiffer.
OECD. (2017). *Systems approaches to public sector challenges: Working with change*. OECD Publishing.
Reeves, T. C. (2006). Design research from a technology perspective. In J. J. H. van den Akker, K. Gravemeijer, S. McKenney, & N. Nieveen (Eds.), *Educational design research* (pp. 52–66). Routledge.
Shute, V. J. (2008). Focus on formative feedback. *Review of Educational Research, 78*(1), 153–189.
Spector, M. (2017). Reflections on educational technology research and development. *Educational Technology Research and Development, 64*, 1415–1423.
Spector, J. M. (2018). Smart learning environments: Potential and pitfalls. In K. Persichitte, A. Suparman, & M. Spector (Eds.), *Educational technology to improve quality and access on a global scale* (pp. 33–42). Springer.
Spector, J. M., & Anderson, T. M. (2000). *Integrated and holistic perspectives on learning, instruction and technology*. Kluwer Academic Publishers.
Sterman, J. D. (2011). Communicating climate change risks in a skeptical world. *Climatic Change, 108*(4), 811–826.
van Merriënboer, J. J., & Kirschner, P. A. (2017). *Ten steps to complex learning: A systematic approach to four-component instructional design*. Routledge.
Wiener, N. (1954). *The Human use of human beings: Cybernetics and society*. Houghton Mifflin.
Wood, D., Bruner, J. S., & Ross, G. (1976). The role of tutoring in problem solving. *Journal of Child Psychology and Psychiatry, 17*(2), 89–100.

The manufacturer's authorised representative in the EU is Springer Nature Customer Service Centre GmbH, Europaplatz 3, 69115 Heidelberg, Germany. If you have any concerns regarding our products, please contact ProductSafety@springernature.com

Printed and bound by CPI Group (UK) Ltd, Croydon, CR0 4YY
30/09/2026
02225261-0003